CHILDHOOD OBESITY: RISK FACTORS, HEALTH EFFECTS AND PREVENTION

PUBLIC HEALTH IN THE 21ST CENTURY

Additional books in this series can be found on Nova's website
under the Series tab.

Additional E-books in this series can be found on Nova's website
under the E-book tab.

METABOLIC DISEASES - LABORATORY AND CLINICAL RESEARCH

Additional books in this series can be found on Nova's website
under the Series tab.

Additional E-books in this series can be found on Nova's website
under the E-book tab.

CHILDHOOD OBESITY: RISK FACTORS, HEALTH EFFECTS AND PREVENTION

CAROL M. SEGEL

EDITOR

Nova Science Publishers, Inc.

New York

For permission to use material from this book please contact us:
Telephone 631-231-7269; Fax 631-231-8175
Web Site: http://www.novapublishers.com

NOTICE TO THE READER

The Publisher has taken reasonable care in the preparation of this book, but makes no expressed or implied warranty of any kind and assumes no responsibility for any errors or omissions. No liability is assumed for incidental or consequential damages in connection with or arising out of information contained in this book. The Publisher shall not be liable for any special, consequential, or exemplary damages resulting, in whole or in part, from the readers' use of, or reliance upon, this material. Any parts of this book based on government reports are so indicated and copyright is claimed for those parts to the extent applicable to compilations of such works.

Independent verification should be sought for any data, advice or recommendations contained in this book. In addition, no responsibility is assumed by the publisher for any injury and/or damage to persons or property arising from any methods, products, instructions, ideas or otherwise contained in this publication.

This publication is designed to provide accurate and authoritative information with regard to the subject matter covered herein. It is sold with the clear understanding that the Publisher is not engaged in rendering legal or any other professional services. If legal or any other expert assistance is required, the services of a competent person should be sought. FROM A DECLARATION OF PARTICIPANTS JOINTLY ADOPTED BY A COMMITTEE OF THE AMERICAN BAR ASSOCIATION AND A COMMITTEE OF PUBLISHERS.

Additional color graphics may be available in the e-book version of this book.

Library of Congress Cataloging-in-Publication Data
Childhood obesity : risk factors, health effects, and prevention / editor,
Carol M. Segel.
p. ; cm.
Includes bibliographical references and index.
ISBN 978-1-61761-982-3 (hardcover)
1. Obesity in children. 2. Obesity in adolescence. I. Segel, Carol M.
[DNLM: 1. Obesity. 2. Adolescent. 3. Child. WD 210]
RJ399.C6C483 2010
618.92'398--dc22
2010036033

Published by Nova Science Publishers, Inc. † *New York*

Contents

Preface

At the beginning of the third millennium, a rising prevalence of overweight and obese children and adolescents were seen in developed as well as developing and threshold countries. According to the WHO, overweight or obesity affects one in ten children or adolescents worldwide. This tendency is a dramatic one because childhood obesity is not only an aesthetic problem which may result in social stigmatization of affected children, but childhood obesity is a multisystem disease with potentially devastating consequences. As with obesity in adults, childhood obesity is acknowledged to be one of the most important risk factors for hypertension and diabetes during childhood and later in life. This book presents current research in the study of childhood obesity, including physical activity promotion programs to thwart childhood obesity; physiological and psychosocial risk factors in childhood obesity; and the ethnic differences in pediatric obesity and metabolic syndrome.

Chapter I - During the past two decades the prevalence of overweight and obesity in children has increased rapidly worldwide. Together with the rising of childhood obesity, some adult-like metabolic disorders, such as the impairment of insulin sensitivity or hypertension, become apparent already among school-aged children. To explain these scaling, factors associated with changes in the social, economic and physical environment together with some genetic background may be of concern. As both a consequence and a contributor to the childhood obesity, the low-level physical activity of children, especially in the westernized societies, warrants attention. Physical activity is embedded to the socioeconomic, cultural, and physical environment of children, and need to be promoted worldwide so that each child could engage in a minimum of 60-min of moderate-to-vigorous physical activity on a daily basis. To this end, it seems crucial to know and handle properly the determinants of the habitual physical activity of children, especially when it is about overweight/obese children.

This chapter is intended to emphasize the value of physical activity in preventing and combating childhood obesity and its metabolic comorbidities. It will be divided into three main sections. The first section discusses metrological aspects of children's physical activity regarding assessment tools and their relevance in pediatric obesity research. The second section is a synthetic state of the art on the relationships between physical activity and childhood obesity and its related diseases. The final section suggests a framework for physical activity promotion programs so as to thwart the high prevalence of pediatric obesity and enhance metabolic profile of children.

Chapter II - *Purpose:* To examine the co-morbidity of the physiological risks tied with psychosocial and environmental factors and present a prospective for prevention of overweight, obesity and related disorders.

Objectives: (1) Describe obesity as a complex phenomenon with multiple and interacting risk factors; (2) Describe the significant increase in sedentary lifestyle coupled with high consumption and availability of calorie dense foods as the catalyst to an obesity epidemic in US; (3) Demonstrate that an understanding of weight gain and weight management with their complexities and multiplier effects is imperative; (4) Review the genetic base for understanding obesity which has advanced in the past decade due to the scientific discovery of several genes that regulate body weight; (5) Demonstrate the environmental contribution to obesity which has also been advanced as society promotes increased calorie consumption and decreased physical activity; (6) Understand the psychosocial factors, such as forced conformity to unrealistic norms, denial of success and access to those who do not fit a standard; and the association of mood, serotonin, and food as medicine to suppress depression which must also be factored in when proposing a novel approach for the prevention of obesity; (7) Demonstrate use of the multiplier approach, where genetics, environmental and psychosocial factors are addressed; (8) Recommend strategies to address the obesity epidemic in US.

Chapter III - Along with the rapid economic and social development during the past three decades, China is experiencing a dramatic epidemiologic transition. Noncommunicable chronic diseases including cardiovascular disease, cancer, and type 2 diabetes mellitus have been becoming the leading cause of death and disability among Chinese people. Obesity plays an important role in the pathogenesis of cardiometabolic disease. There is a growing body of evidence suggesting that overweight or obesity in childhood are associated with an increased risk of being overweight or obese and having obesity related metabolic disorders during adulthood. Few studies have systematically assessed changes in the prevalence of overweight and obesity, the health consequences, and related risk factors among Chinese children and adolescents. In this review, the authors attempt to summarize: 1) the prevalence of overweight and obesity and its time trends among Chinese children younger than 18 years of age; 2) overweight and obesity related metabolic abnormalities; and 3) associated risk factors.

Chapter IV - Childhood obesity is strongly linked to future diseases in later life, including cardiovascular disease (CVD) and Type 2 diabetes mellitus (T2DM). Prediction of these future diseases is assisted by diagnosis of the metabolic syndrome (MetS), a cluster of cardiovascular risk factors linked to insulin resistance. African Americans have higher rates of insulin resistance, T2DM and death from CVD but paradoxically have lower rates of MetS diagnosis. This is largely because of ethnic differences in triglyceride levels. If we are to accurately identify obese children who are in highest need of weight loss intervention, new risk predictors are needed that perform well among all ethnicities.

Chapter V - Obesity rates are increasing worldwide at an alarming rate. Today more than 1.1 billion adults worldwide are classified as overweight, 312 million of them as obese. Especially dramatic is the situation among children and adolescents. According to the WHO overweight or obesity affects one in ten children or adolescents worldwide. In Europe one in five children can be classified as overweight or obese. Although overweight and obesity are becoming increasingly prevalent nearly all over the Industrialised world special risk groups can be identified. In the United States as well as in Europe migrant status seems to increase the risk for developing overweight or obesity during childhood an adolescence dramatically.

Migration per se does not lead to increased obesity rates but migrant status is still associated with a low socioeconomic position, adverse nutritional habits and a low level of physical activity. Additionally religious and social restrictions increase the problem. In this paper the impact of migration on obesity prevalence among Turkish children living in Vienna Austria, is focused on. Beside methodological problems the impact of migration on behavioural and social factors during childhood is discussed. Within the European Union the majority of migrants suffering from excessive body weight originate from Mediterranean countries, but also from the Near and the Middle East. Among this subgroup an effective prevention of childhood obesity seems to be extraordinary difficult because cultural and religious norms make any prevention concept containing changes in eating behaviour and physical activity patterns difficult.

Chapter VI - All over the world obesity represents a social and medical emergency. The obesity epidemic is particularly alarming in young age, as overweight or obese children and teen-agers are likely to become obese adults. Overweight and obesity in childhood and in adolescence can be ascribed to genetic factors as well as to socio-environmental factors peculiar to the principal contexts of life. All over the world, public institutions are taking action to break this vicious circle. Also in Italy, national, regional and local projects have been developed to favor healthy lifestyles in children, following a prevention model integrating 'macro' actions (of political and social nature) with educational interventions directed to the 'micro' environment of individual subjects.

In this report, the authors revise the literature in the area of healthy life-style promotion in childhood, focusing on the results of a few Italian programs on the nutritional and physical habits of children. The main purpose was to demonstrate whether interventions aimed at increasing physical activity could produce significant changes in children's behavior, as well as awareness about the needs of healthy choices in the family.

Chapter VII - Advances in portable electronic technologies have created opportunities for the real-time assessment of children's physical activity and eating behaviors in naturalistic situations. Mobile phones or PDA's can be used to record electronic surveys, take photographs, or indicate geographic locations of children's behaviors. Unlike self-report instruments, which are prone to recall errors and biases, real-time data capture (RTDC) methods can assess behaviors as they occur. In addition, these strategies are able to provide contextual information about physical activity and healthy eating such as where and with whom the behaviors are taking place; and how children feel before, during, and after these activities. This commentary will describe how RTDC methods can enhance our understanding of factors influencing children's physical activity and eating behaviors. In particular, it will discuss the potential to advance research pertaining to the following questions: (1) How frequently, when, what amount, what intensity, what duration, and what type of food or activity was eaten or performed? (2) Where and with whom do children engage in physical activity and eat, and do these patterns differ according to demographic (e.g., sex, age, ethnic, income) and temporal (e.g., time of day, day of the week, seasonal) characteristics?; (3) How do children's physical activity levels (e.g., intensity, duration) and eating patterns (e.g., amount/content of food) differ across physical and/or social contexts?; (4) To what extent do mood, stress, and psychosocial factors serve as time-related antecedents and consequences to children's physical activity and eating episodes?; and (5) Are patterns of within-daily variability in children's physical activity and eating behaviors related to health outcomes such as body weight, insulin dependence, and the metabolic syndrome? The commentary will also discuss practical and economic challenges associated with employing RTDC methodologies in research studies with children. It will conclude by addressing how these innovative research strategies can inform the design of programs and policies to prevent and treat childhood obesity.

Chapter VIII - Childhood obesity is a multifactorial disease resulting from an imbalanced energy intake and expenditure. The results from some investigations demonstrated that children are more likely to be active during school recess time; for that reason school playgrounds offer a sustainable context to increase vigorous physical activity in a way to counterbalance children's inactivity and to decrease obesity. The purpose of this study is to explore the association among obesity and gender, school year, physical activity, time spent going to school and how, areas and equipment of the schoolyard.

The sample included 975 children (505 girls and 470 boys) (age 8,28 + 1,23) from elementary school. Obesity was estimated by BMI and the cut off points of Cole et al. (2000). A questionnaire was completed by parents to provide information about age, school year, gender, time spent going to school and how, as well as the practise of physical activity. The areas (total and children per m2) from the schoolyards were calculated using the AutoCAD software. The schoolyards were characterized according to the equipments into four categories: non equipment; playground equipment; sport equipment and playground+sport equipment. The logistic regression was used to estimate the magnitude of association among overweight plus obesity and independent variables.

The results from multinominal logistic regression were only significant for time spent going to school (OR=0,612; 95% CI 0,437 – 0,856), total area (OR=1,756; 95% CI 1,036 – 2,975) and type of equipment - non equipment (OR=1,723; 95% CI 1,084 – 2,740) and playground equipment (OR= 2,048; 95% CI 1,063-3,947). The odds ratio of being obese was 0,612 times less for children that spend less than 10 minutes going to school. On the other hand, schoolyard area is a risk factor for obesity, 1,756 times more for children that play in a schoolyard with an area less than 1000 square meters. Also the absence of physical structures and the type of equipment are risk factors. The schoolyards that don't have any kind of equipment have a risk factor of 1,723 more to have obese children and those with playground equipment have a 2,048 higher risk than schools with a variety of equipment.

The results from this study highlighted the importance of schoolyards designed in a way so that the increase of physical activity and to reduce obesity. To promote healthy lifestyles, the authors suggest a multidisciplinary team to design playgrounds, taking into account the area and the physical structures.

Chapter IX - Excess weight population is increasing in developed and developing countries and are becoming a serious health concern. The Spanish Society for the Study of Obesity (SEEDO) indicates the importance of addressing it on a high-priority basis. In this article, the authors propose a mathematical model of epidemiological type to predict the incidence of excess weight in 24- to 65-year-old females and males in the region of Valencia (Spain) in the coming years. Due to the intrinsic uncertainty of the statistical data used, the authors also apply the Latin Hypercube Sampling (LHS) to the model in order to obtain predictions over the next few years providing 90% confidence intervals.

Chapter X - This study examines leading childhood obesity journals' instructions to authors regarding policy-related research and implications. A systematic approach using five data sources was used to determine 50 leading childhood obesity journals. The following information was obtained from each journal: aims and scope; instructions to authors; types of articles published; and, each type of articles' format and abstract structure. Fifteen of the 50 journals (30%) included policy in their aims and scope. Eighteen journals (36%) explicitly instructed authors on a type of article where policy-related information could be published. Only three journals (6%) explicitly indicated a potential policy outlet within their research article format and one journal provided a potential policy outlet within their abstract structure. Opportunity exists for the development of more explicit editorial policies and best practices on how researchers incorporate policy components in their study design and articulate policy implications in their publications.

Chapter XI - The purpose of this chapter was to review nutrition education in school-based childhood and adolescent obesity prevention programs. Problems of overweight and obesity have reached epidemic proportions in children and adolescents. With the large number of children and adolescents affected by overweight and obesity, it has been recognized as a critical public health threat, and prevention should become a priority. Schools have been considered an important venue for the delivery of health education programs since a majority of the nation's youth can be reached. Implementing a program in the school does not necessarily correspond to success, and therefore common components among successful nutrition education programs were examined. Providing age-appropriate, culturally sensitive instruction and materials were connected to the success of the programs. Success was also found when the nutrition education program was relevant to the student, and information was applicable to their everyday life. It was also critical to go beyond dissemination of knowledge and develop attitudes, skills, and behaviors to adopt, maintain, and enjoy healthy nutrition habits. Finally, recommendations were made to enhance the effectiveness of future school-based nutrition education programs.

Chapter XII - Very little is known currently about the pattern of risk for early childhood overweight and obesity in the least developed countries, where child under-nutrition remains very common and a pressing concern. The authors use standardized anthropometric and interview data pertaining to seventeen nationally-representative samples of 37,714 children aged between 30 and 60 months to model that risk. The authors particularly consider the possible roles of changing social and economic status of households and urban residence, and take into account such factors as variations in family size, in maternal nutritional status, and children's histories of under-nutrition (observed as growth stunting). The relationships among these variables are quite different across world regions except for mothers' overweight status, which was a strong predictor in all. In sub-Saharan Africa, overweight children are extremely rare and the only strong predictor is having a mother who is overweight. In Northern Africa urban residence is a risk factor. In the Americas, increasing wealth and social status of households raises risk substantially. Stunting places children in Africa, but not the Americas, at significantly increased risk of being overweight, and in northern Africa this effect is particularly pronounced in cities. The authors find every indication in these trends that child obesity and overweight might very quickly emerge as the modal nutritional status of children worldwide. The model suggests that childhood overweight in many ways embodies relative poverty as national wealth rises, just as child stunting reflects the conditions of absolute poverty. As economic growth accelerates in the poorer countries, the least advantaged sectors of their populations can remain absolutely poor while their relative poverty also increases. This eans that risk for childhood obesity will grow, and probably rapidly, and it will increasingly co-exist with and so be intensified by under-nutrition.

In: Childhood Obesity: Risk Factors, Health Effects...
Editor: Carol M. Segel

ISBN: 978-1-61761-982-3
©2011 Nova Science Publishers, Inc.

Chapter I

The Importance of Physical Activity in a Lofty Pediatric Obesity Context

Benjamin C. Guinhouya
University Lille-Northern France, France

Abstract

During the past two decades the prevalence of overweight and obesity in children has increased rapidly worldwide. Together with the rising of childhood obesity, some adult-like metabolic disorders, such as the impairment of insulin sensitivity or hypertension, become apparent already among school-aged children. To explain these scaling, factors associated with changes in the social, economic and physical environment together with some genetic background may be of concern. As both a consequence and a contributor to the childhood obesity, the low-level physical activity of children, especially in the westernized societies, warrants attention. Physical activity is embedded to the socioeconomic, cultural, and physical environment of children, and need to be promoted worldwide so that each child could engage in a minimum of 60-min of moderate-to-vigorous physical activity on a daily basis. To this end, it seems crucial to know and handle properly the determinants of the habitual physical activity of children, especially when it is about overweight/obese children.

This chapter is intended to emphasize the value of physical activity in preventing and combating childhood obesity and its metabolic comorbidities. It will be divided into three main sections. The first section discusses metrological aspects of children's physical activity regarding assessment tools and their relevance in pediatric obesity research. The second section is a synthetic state of the art on the relationships between physical activity and childhood obesity and its related diseases. The final section suggests a framework for physical activity promotion programs so as to thwart the high prevalence of pediatric obesity and enhance metabolic profile of children.

Introduction

In a period that can be considered as an epidemiological transition [1], overweight and obesity appears as the fifth major health risk of death worldwide [2]. Indeed, up to the 2000s where stabilization seems to arise, the prevalence of overweight and obesity has dramatically increased in many countries in the last two decades. This can be exemplified by US data where overweight and obesity has increased from 32% and 13% in 1960-1962 to 68% and 34% in 2007-2008 [3]. This trend, albeit at a lesser extend, was also found in Europe, Asia, South America, and even in Africa [4, 5]. In parallel to the increase of obesity in adults, pediatric population expresses more than ever a high body mass index (BMI), with more than one-third of US children being overweight or obese [6, 7], 20% in Europe, and 15% in the Near/Middle East [6]. In the US and Europe as well, the number of overweight and obese children has doubled, and the number of obese adolescents has tripled over the past 20 years [8]. Although a plateauing trend of the pediatric obesity has been found in different geographical areas [7, 9], this prevalence still being too high, and actions need to be intensified to curve definitely such a propensity.

Excess weight during growing years is of great importance for at least three main reasons. First, it constitutes a risk of overweight and obesity during adult life. Second, childhood obesity correlates with adulthood risk factors for common chronic diseases such as diabetes, cardiovascular disease, and other morbidities. Third, obese children already undergo many health adverse effects including orthopedic problems, sleep apnea, impaired quality of life, psychological or mental disorders, and some cardiometabolic abnormalities. One of the most important obesity-related metabolic disorders is currently known as the metabolic syndrome (MetS). MetS is assumed as the clustering of at least three risk factors that fit the following criteria: hypertriglyceridemia, hypertension, abdominal obesity, low concentration of high-density lipoprotein HDL-cholesterol (HDL-C), and high fasting blood glucose. Overt MetS during childhood is indicative of an increased risk of morbi-mortality from type 2 diabetes, cardiovascular diseases (CVD), and a loss of functional capacities throughout life. This observation highlights the magnitude of future health and economic consequences at the individual and society levels. Although, various definitions have been used for the pediatric MetS, the International Diabetes Federation (IDF) has recently recommended for the diagnosis of the syndrome in children to identify a central obesity plus at least any two others of the aforementioned criteria [10]. From the available literature, the prevalence of MetS may go up to 60% in obese children, while it occurs in less than 1% of normal weight children [11, 12]. Because obesity is intimately associated with the pediatric MetS [13], which may track into adulthood including its clinical complications [14, 15] the actual high proportion of overweight/obese children should be considered symptomatic of a larger decline in the overall health of the population [16].

Simplistically, obesity reflects a 'loss of control' of energy balance such that over time, the excess energy is stored as fat. The mechanisms underlying this 'loss of control' are complex, multi-faceted and not yet well-understood. Very likely, it may be that a genetic (surely, an epigenetic) background combined with a 'toxic' environment that furthers inappropriate behaviors, is at work in the high prevalence of pediatric obesity. As a matter of fact Bouchard [17] assured that predisposing genes can be enhanced or diminished by exposure to relevant behaviors. One of the behaviors in question is the habitual physical activity of individuals. Even if it should be acknowledged that the relative contribution of energy intake *versus* energy expenditure to obesity epidemic is a source of continuing debate, it has been empirically found that physical activity declined more greatly than food consumption over time [18-20]. On principle, energy expenditure drives energy intake rather than the reverse [21]. It seems that the subsistence efficiency, as defined by the ratio of energy intake and expenditure has increased from 3:1 among ancestral humans to 7:1 in the modern life, due mainly to a greater decrease in energy expenditure [22]. Thus, the deficit in the physical activity-related energy expenditure unit is to be re-established in an environment conducive to sedentary pursuits. For children and youth, the current public health goal regarding physical activity is to spend at least 60 min per day in a moderate-to-vigorous physical activity (MVPA) [23, 24]; an activity intensity that is equivalent to brisk walking. Some specifications have been developed for infants and toddlers [25-27]. Compelling data however suggest that youth in much of Western world have relatively low levels of MVPA, and spend a large percentage of their time in sedentary pursuits [28], whatever the tool used to assess their physical activity. Although sedentary behavior might not be a subject of discussion in this chapter in spite of its putative implication to the pediatric obesity, the same authors underscored the little change in sedentary behavior *per se* over time, whereas there is a high prevalence of children who do not meet recommendations for physical activity [28].

Because physical activity measurement issues may hinder the exploration of the relationship between physical activity and health outcomes, and especially adiposity of children, the first section is devoted to clarify tools that should be used in epidemiological studies of pediatric obesity. The second section reviews the current knowledge about the relationships about children's physical activity and obesity and its related metabolic disorders. The third section suggests a model for physical activity promotion in the pediatric population so as to prevent obesity and its related comordities in this age group.

1. Measurement and Surveillance of Physical Activity in the Pediatric Obesity Research

Physical activity is a behavior that theoretically includes any type of movement from fidgeting to the involvement in extreme and competitive sports. It was defined as any bodily movement produces by skeletal muscles, resulting in energy expenditure [29]. As a behavior, physical activity may not be reduced to only sports, and is embedded in the social and cultural context of individuals. Furthermore, although physical activity can be considered as the portion of daily energy expenditure that is the most amenable to change, it consists of two parts – movement and energy expenditure (the consequence of that movement) – that may be measured. As such the two terms should not be used synonymously.

The habitual physical activity of children, as a latent time series of activity type and intensity [30], comprises all activities performed in the course of a day in an organized setting (e.g., school, sports club, leisure centre), during transport/commuting, and while playing with peers. In each of these environments, important dimensions need to be characterized. These include intensity, frequency, and duration, which put together determine the overall activity level of a child. An additional dimension of physical activity is the type or mode such as walking, running, and so on.

Ideally, a physical activity measurement tool should provide valid and reliable information of all dimensions of this behavior in all of its domains. This makes challenging the measurement of physical activity in the general population. There is a further concern with this measurement among children given the intermittent and sporadic nature of their spontaneous physical activity [31] when compared to adults. Bailey et al. [31] were the first to show that about 95% of the activity bouts of children do not exceed 15-s, and only 0.1% of their active period lasts more than one minute. This has important implications for all aspects of measurement, processing, and interpretation of physical activity data in youth, including sampling frequency.

To date, a wide range of methods has been used to measure physical activity in children and adolescents. These include self-report methods (such as questionnaires and activity logs/diaries), direct observation, and more technology-based tools such as double labelled water (DLW), heart rate (HR) monitors, and motion sensors (e.g., pedometers and accelerometers). This section summarizes specific aspects (principle, advantages and disadvantages) of the different assessment tools, and discusses the relevance of their use in the pediatric obesity research.

1.1. Self-Report

Self-report instruments include diaries, questionnaires and proxy or parental reports of children's activity. These techniques are based on the recall of physical activities in a more or less recent past (recall the activities of the last 3 or 7 days or even from the previous month). The required time to complete questionnaires vary from some minutes to several hours depending on the objectives pursued, the comprehensibility of questions and the availability of respondent.

The main advantages of self-report measures are their ease of administration, the ability to characterize activity historically, and their low cost. The ability to record activity types and context in which this occurs are also advantages. As a consequent, self-reports have been commonly used in epidemiological research and surveillance. However, they have considerable limitations, especially with children. These include difficulties to account for the sporadic nature of children's activity (concern about the intensity and duration of bouts of activity) and the lack of objectivity, which cannot exclude information biases, likely due to exaggerated perceptions (by either child or parent) and the limited cognitive ability of younger children. Thus, these subjective measures appear to quantify the perception of physical activity rather than physical activity *per se* [32, 33]. As such, children or their proxy-reporters may tend to overestimate children's activity level, given the higher estimates obtained from self-report methods as compared to other instruments [34].

1.2. Direct Observation

Observation typically involves observing a child at home or school for extended periods of time and recording into either a laptop computer or coding form an instantaneous rating of child's activity level [35]. One major advantage of the direct observation is that it allows identifying the proximal determinants (contextual information) of physical activity in children: behavioral cues, environmental conditions, presence of partners, availability of equipment and toys. The direct observation of children's activity, which is often used as a validation criterion, may be more suitable for the assessment of physical activity in controlled situations. On the other hand, this technique is expensive, laborious, time consuming and requires an extreme diligence from observers [36]. Finally, subjects' reactivity to observers is another matter of concern for a method that might appear intrusive for participants. For all these reasons, direct observation may be of limited interest in epidemiological studies.

1.3. Double Labeled Water (DLW)

The DLW method is viewed as the gold standard for the measurement of energy expenditure in free-living conditions, and is often used as a validation criterion of most activity monitors. In the DLW method a body-size dependent dose of the isotopes ^2H_2O and $H_2{}^{18}O_2$ is ingested. Hydrogen and oxygen are involved in all metabolisms and are subsequently excreted from the body in a rate in proportion to the metabolic rate. The oxygen isotope is excreted as $C^{18}O_2$ and $H_2{}^{18}O$ but the hydrogen isotope as ^2H_2O. The difference in the rate of loss (obtained through urine, sweat, evaporative losses) between the two isotopes is a function of the rate of carbon dioxide production, an indicator of the rate of energy production over time [37]. A major limitation associated with DLW is its excessive cost. Furthermore, it does not provide estimates of energy over different intensity levels.

1.4. Heart Rate Monitors

Heart rate monitors use the electrical signal from the heart to measure each heartbeat. Physical activity is not directly measured in this way, but – perhaps more importantly – the relative stress being placed upon the cardiopulmonary system by the activity is monitored. The device consists of lightweight transmitter, which is fixed to the chest either by a belt or with electrodes, and a receiver/microcomputer, which is worn as a watch on the wrist. Heart rate monitors can be used to assess the frequency, intensity and duration of physical activity, based on the relationship between heart rate and oxygen uptake. While this relationship may be linear at moderate intensities, this is not so at higher and lower intensity levels. This may be a source of error since most children spend a large proportion of their day in sedentary and light activity [38]. Furthermore, heart rate signals may be influenced by emotional stress or type of activity undertaken and fitness levels. This means that heart rate monitors should be calibrated for each subject by adjusting for fitness level. The heart rate monitors can be subjected to interference problems, loss of signals due to unexpected interruptions that make the collection and processing of data, at best, circumspect. Finally, wearing a strap/belt on the chest can be a real barrier - the point of jeopardizing the compliance of children to the protocol as the data recording time lengthens. All these drawbacks may limit the use of heart rate monitors in large studies.

1.5. Motion Sensors

Pedometers

Pedometers, which estimate the number of steps taken over a given period, are the simplest method to assessing ambulatory activities. Most pedometers consist of a horizontal spring-suspended lever arm that moves with the vertical acceleration of the hips during ambulation [39]. They are relatively cheap and usually well tolerated by subjects and can therefore be used in epidemiological studies. Pedometers can be relevantly used to deliver public health messages, such as *"Just take 15000 steps per day for your health!"* or as an activity promotional tool in the way of an open-loop feedback on physical activity [40]. Interestingly, steps obtained from pedometers are increasingly matched with accelerometer data and translated in order to offer pedometer-based physical activity recommendations for children [41, 42]. As such, pedometers may have a great interest in epidemiology, for measurement and surveillance purposes. On the other hand, pedometers have the basic limitation of motion sensors, in that they are insensitive to some forms of movement. In addition, these devices do not in any way allow detecting specific intensity categories (e.g., light, moderate, or vigorous). Finally, some pedometers do not possess real-time data storage and downloading capabilities.

Accelerometers

Accelerometers have become increasingly popular as an objective method of assessing physical activity. There are a number of commercially available monitors that work using the same principles, normally using electro-mechanical piezoelectric levers to detect acceleration. More details about the technical properties of accelerometers have been provided by Chen and Bassett [43]. As a function of the number of plane the accelerometer can detect, there are uniaxial (e.g., The Actigraph LLC, Pensacola, FL, USA) for in single plane detection, bi-directional (e.g. Biotrainer IM Systems, Baltimore, MD, USA). Other accelerometers are triaxial (e.g., Tritrac-R3D, Hemokinetics, Madison, WI, USA) and are able to detect acceleration in three planes, while some others are omni-directional (e.g., Actical, Mini-Mitter, Bend, OR, USA). Accelerometers can be placed at various sites on the body, including the wrist, ankle, or hip. In epidemiological studies the hip wearing is very frequent.

The Actigraph monitor, produced by Actigraph, LLC, is the most widely used accelerometer. This type of accelerometer has been found to have the best the clinimetric quality over other commercial accelerometers [44-46]. The Actigraph integrates accelerations/decelerations in the vertical plane via a piezoelectric plate. Acceleration detection ranges from 0.05 to up to 2.50 g in magnitude and the frequency responses range from 0.25 to 2.5 Hz, so that motion outside normal human movement is rejected by a filtered bandpass. The acceleration-deceleration signal is digitized by an analog-to-digital converter and numerically integrated over a user-defined sampling interval or 'epoch". The signal is sampled 10 times per second and the data sorted into epochs and stored in the internal memory; then the integrator is reset to zero. To begin data collection, the monitor is initialized through a compatible personal computer. A real-time internal clock allows the researcher to begin collecting at the desired time. The output from the Actigraph is expressed in "counts" per epoch. "Counts" represent the summed amount and magnitude of acceleration during each epoch. The epoch can vary from 1-sec to several minutes. One-minute epochs have generally been used in field studies, and this allows approximately 22 days of recording with the model 7164, and >300 days with the model GT1M. There is a newer model, the GT3X, which is deemed to capture motion in up to the three dimensions. Current models of the Actigraph accelerometer (i.e., GT1M and GT3X) are waterproof and can capture movement during swimming. The internal clock in the Actigraph allows time and duration as well as intensity of activity to be monitored, thus daily patterns of physical activity can be described. One main drawback of accelerometers in general, and particularly the Actigraph accelerometers, is the failure to detect all types of movement (e.g. cycling, swimming for the Actigraph model 7164). In addition, the high volume of data outputted by many instruments and the need to use sophisticated, researcher-designed post-processing programs to screen data for factors such as minimum wear time, off-body periods [47] is time consuming and may cause enormous burden to researcher, and postpone the computation of physical activity outcomes. Finally, despite recent advances on the topic [33, 48-50], the gap is not yet closed regarding the cut-off point that should be used to define the MVPA of children [51].

The choice of a method is justified by several criteria including its validity, reliability, ease of administration, but also the opportunity to use it or not in children of a certain age. The monitoring technique must be socially acceptable; it should not burden the child with cumbersome equipment, and should have a minimal influence of the child's normal physical activity patterns. Apart from methodological criteria and ethics, should be added the cost of the technique. For each of the category of technique some level of validity has been found in the literature. Thus, each of them may be useful in the clinical setting subject to the recognition of their drawbacks as mentioned above, and the control of their specific biases. For example, data obtained from proxy- or self-report should be used cautiously because of exaggerated perception of the activity level of children [32, 33, 52]. However, in epidemiological research, other tradeoffs need to be considered, particularly regarding cost-effectiveness. As mentioned earlier, none of the existing methods fully meets all the criteria, each with advantages and disadvantages. It should nevertheless be stressed that because of their cost and therefore the impossibility to use them for monitoring and epidemiological surveillance, direct observation and calorimetry are generally reserved for the validation of other methods. Apart from their inherent limitations, the heart rate monitors are prone to discomfort, which may be greater among overweight and obese. The wearing of a strap/belt on the chest may hinder adherence to protocols by overweight/obese children as compared to their leaner counterparts. Thus, the remaining two choices for epidemiological studies are self-report methods and activity monitors.

Although self-report methods are often the measurement of choice, particularly in large-scale epidemiological investigations, they are frequently prone to error that may attenuate the association between physical activity and an outcome of interest toward the null hypothesis of no association [53]. Self-report measures of physical activity in children produce estimated activity levels that are consistently higher than those reported with other instruments [34, 54]. For instance, Basterfield et al. [55] recently found that the surveillance system of the physical activity of children in the UK may be flawed due to the use of parent-reported data. It was found that 80% of parents of inactive children considered their children to be sufficiently active, while 40% of inactive children overestimated their own physical activity level [32]. Finally, it was reported that overweight children may perceive a given intensity of physical activity to be more strenuous than their normal-weight peers, particularly at higher levels of intensity [56-58]. This can explain why overweight children tend to over-report their MVPA as compared to normal-weight girls [59]. Thus, cautious may be required in using proxy- or self-reported physical activity data for the monitoring and surveillance of activity in the pediatric population, especially when overweight and obesity are the outcomes of interest.

The use of motion sensors, and especially of accelerometers, was at one time confined to use in small groups but are now being more widely use in large epidemiological studies. This is exemplified by the increasing number of large-scale international and national databases that include accelerometry-based physical activity data of children and adolescents. As examples in Europe, the European Youth Heart Study (EYHS, which includes >2,000 children aged 9-15 year-old) [60, 61], the Avon Longitudinal Study of Parents and Children (ALSPAC, which includes >5000 children of 11 years) [38]. In the US, there is the National Health and Nutrition Examination (NHANES, which contains accelerometer data of >6,000 participants including >1,500 children of 6-19 years) [62]. Alternatively, pedometers have been successfully used in about 43 studies with sufficient sample sizes for epidemiological analyses [63]. In Canada particularly, a pedometer-based surveillance system, the Canadian Physical Activity Levels among Youth (CANPLAY, which includes >11,000 children aged 5-19 year-old), has also been successfully carried out [64, 65].These research endeavors are likely due to the growing prevalence of pediatric obesity and the willingness of governments and funders to invest in physical activity research.

The adoption of motion sensors, and mainly accelerometers, into physical activity research among youth has enhanced the measurement of physical activity. This has indeed enhanced our understanding of the relationship between physical activity and obesity, as well as a range of other health outcomes. Although accelerometers, indubitably have advantages over subjective instruments (e.g. self report) they still have shortcomings and there are some unresolved issues in the methodology and interpretation of data [66]. Nonetheless, accelerometers have greatly helped in solving accuracy and reliability issues related to self-report methods, and allow unifying strategies in physical activity measurement. Even if studies are needed to increase the standardization in the use of accelerometers, these devices seem to be feasible, valid and reliable enough to be recommended for epidemiological research in pediatric obesity. Alternatively, pedometers may be encouraging tradeoff for surveillance and physical activity promotion purposes.

2. Current Evidence on the Effect of Physical Activity on Childhood Obesity and its Metabolic-Related Disorders

A recent systematic review summarized the health benefits of physical activity in school-aged children and youth [67]. From this review, Janssen and LeBlanc [67] found that the more physical activity, the greater the health benefit. The authors have come to a conclusion that substantiates the current recommendations of physical activity for school-aged children [23, 24]; that is, a minimum of 60 min *per* day of at least a moderate-intensity physical activity. Nonetheless, a low-intensity physical activity still remains better than a chronic sedentary lifestyle [68]. The conceptual inverse relationship between activity levels and body fat of children has long been obscured by the variety of methods used to assess physical activity, including the frequent use of self-report techniques among children. These relationships are however enhanced with the use of motion sensors [69]. Thus, studies using self- or proxy-reported physical activity measures among children tended to report weak to modest relationships between physical activity and overweight/obesity, with many risk estimates being non-significant [67]. On the other hand, a growing number of recent studies, using accelerometry-based data in large cohorts, consistently reported an inverse relationship between physical activity and indicators of childhood obesity [70-72]. Nevertheless, these studies are mostly cross-sectional observation studies, and no cause-and-effect relationship might be drawn from such findings. It still remains that the median odds ratio for overweight/obesity in the least active group relative to the most active group were 1.33 and 3.79 for self-report and objective methods, respectively [67].

The few existing longitudinal studies [73, 74] lend support a causal relationship between physical activity and body fat in the pediatric population. In a one-year prospective study among 10-15 year-old children an increase in physical activity was found to be associated with a decline in BMI among girls and overweight boys [73]. The authors speculated that the extent of this decline may be weakened due to the use of self-reported data of physical activity. Using Caltrac accelerometers to assess physical activity during a longitudinal study of 8 years, Moore et al. [74] came to the conclusion that active children have much less body fat by the time of early adolescence than those who were less active. These two longitudinal studies confirm the protective effects of physical activity on the long term change in body fat during childhood and adolescence. More recently, not only the protective effects of physical activity have been confirmed, but it has been also found that physical activity at an early age predicted later fat mass even after adjustment for concurrent physical activity [75].

Experimental and quasi-experimental studies using physical activity alone as exposure and adiposity as the outcome of interest among children are rare. Furthermore, the primary aim of most of these intervention studies was to improve other health measures (e.g., blood lipids, insulin resistance, and bone density) and not obesity measures *per se* [67]. It has been reported that half of the exercise interventions, which were aerobic in nature significant changes in measures of BMI, total fat, and/or abdominal fat in response to training. Only some studies that used other training modalities (e.g., resistance training, circuit training, pilates, jumping exercises) observed significant improvements in measures of total fat, abdominal fat, or BMI in response to training. Nevertheless, the effect sizes, even for the studies that found significant improvements, tended to be small for both aerobic and resistance exercises-based programs (<0.50) [67].

It should however be acknowledged that few of these interventions were based on a theoretical framework; most of them were structured or exercise-based interventions, and did not promote long term lifestyle changes, and they scarcely involved several actors at the same time (i.e families/child day-care centres, schools, and community together) or used a multi-sectoral approach (i.e., home, medical consulting room, neighbourhood, and playground). In addition, in most interventions physical activity has been evaluated with self-report methods, which data may attenuate effects toward the null hypothesis of no effect.

Physical activity also appears to be one of the most effective preventive and treatment options in the pediatric MetS, and a number of physiological explanations have been provided regarding the mechanisms through which physical activity may positively impact the Mets or its components [76-78]. Thus, it has been suggested that physical activity can directly influence insulin resistance while causing, *inter alia*, short-term activation of the glucose transporter (GLUT-4) receptors. This occurs through lowering the circulating insulin level by increasing the production of anti-inflammatory factors, such as the adiponectin, or by inducing an increase in the quantity of oxidative and insulin-sensitive fibres. These effects can be direct or mediated by an increase in physical fitness (cardiorespiratory fitness and strength) or by a decrease of adiposity (which may occur in direct correlation to an increased circulating concentration of adiponectin). Further works may straighten these mechanisms. Meanwhile, a growing number of studies lend support a negative relationship between different forms of physical activity and the infantile MetS or its components [79]. The main target of physical activity and exercise programs is insulin resistance [80], although regular physical activity can also counteract the syndrome's various biological risk factors, such as blood pressure [81-84] and lipidemia [81, 85]. These effects can be obtained through structured exercise programs and free-living physical activity as well.

Concerning structured exercise programme, it has been found that aerobic exercise and well-supervised resistance training among children have positive effects on insulin resistance and various components of infantile MetS. Nassis et al. [86] found an improvement of insulin sensitivity in obese children after partaking in a 12-week exercise training. The concentration of circulating hormones was not modified, but it produced a 12% increase cardiorespiratory fitness. Increased cardiorespiratory fitness, achievable through aerobic exercise, may protect against infantile MetS even in the presence of excess weight. Likewise, some resistance exercises have been found to increase muscular strength and protect against a overly high insulin resistance, regardless of the increase in cardiorespiratory fitness [87]. Finally, combining aerobic and resistance exercises could be used to improve the insulin sensitivity of children [88, 89].

Apart from the positive effects (acute or in the medium-term) of structured exercises on insulin sensitivity and MetS, some cross-sectional and longitudinal data illustrate the significance of lifestyle physical activity of children in preventing MetS. Consistent negative relationships have been found between insulin resistance, MetS or their indicators, and the spontaneous activity of children measured using a questionnaire [90, 91] or objectively through accelerometry [81, 92-94]. Commonly, this relationship is independent of other factors (*e.g.*, body composition, physical fitness) [90, 92]. Nonetheless, there may be a longitudinal joint increase in the central adiposity and insulin resistance of children when physical activity decreased [77]. Moreover, it has been suggested that the beneficial effects of physical activity are better in children whose cardiorespiratory fitness is initially weak [92]. Mechanisms similar to those mentioned above may explain these variations. However, it is important to acknowledge that these positive effects are evident only if the intensity of daily living activities sufficiently stimulates the skeletal muscles or adipose tissue, producing appropriate physiological adaptations likely to improve insulin activity. In this regard, some recent data suggest that 30-40 min of a moderate activity per day can reduce MetS risk by one-third [93]. Consequently, it is an overwhelming public health challenge to help children make optimal use of their available opportunities for activity to counter obesity and its comorbidities.

There remains a lack of prospective longitudinal and randomized controlled trials to ascertain the causal and dose-response relationships between children's physical activity and obesity and its related-abnormalities in the pediatric population. This is to be viewed as an important future research area. More emphasis should be put on interventions that use objective measurements of physical activity among children. There is also a need for theory-based multilevel interventions that implicate together families, schools and the community, including a clear policy for environmental changes.

3. How to Promote Physical Activity Among Overweight and Obese Children!

Physical activity needs to be integrated or re-introduced into the daily life of overweight and obese children and adolescents because of their relatively low level of physical activity when compared to their leaner peers. Nonetheless, McMurray et al. [59] pinpointed that the relationship between the decline in physical activity and excess weight may be an iterative loop: less physical activity may cause weight gain due to a positive energy balance; and weight excess may contribute to real or perceived barriers to physical activity, leading to additional weight gain. There is probably the installation of a "vicious circle" that measures to promote physical activity among obese children should try to break or turn into a "virtuous circle" where an increased activity level should be exercised. In this scheme, health promoters should consider a new interventional paradigm (Figure 1).

Meanwhile, both motivation and skills are required for a sustained behavior change. The involvement of overweight and obese children in a long-term physical activity program requires that the program be attractive and fun. Biological data on physical activity suggest that children's motivation for physical activity may be influenced by neurological factors such as increased dopamine sensitivity [95], and especially the development of the opioid system involved in pain reduction [96]. This implies that an increase in physical activity may increase the threshold of pain sensitivity resulting in a better copying of the demanding nature of effort among obese children who tend to rate, at a higher level, their perceived exertion as compared to their leaner peers.

It seems that obesity is also reflected by a metabolic transfer, with a high prevalence of fast-twitch fibers among obese [97]. Thus, to attract obese children to physical activity programs, it would be useful to regularly introduce, during organized sessions, resistance exercises for which they are most likely to do well in the beginning. This will produce the dual effect of increasing their physical fitness and sustain their self-efficacy, self-esteem [98], and therefore enhance their motivation. In addition, resistance exercises have been found to improve insulin sensitivity of children [99]. As a consequence, aside the behavioral modification strategy that is required to enhance the empowerment and the control of the obese child and his/her family over the activity behavior, suitable community interventions that are supervised by physical activity specialists can be implemented to support this behavioral approach.

3.1. Physical Activity Promotion among Obese Children: Paradigm and Actors' Role

Obesity should be viewed as a societal rather than only a medical condition. As such, several actors and multiple sectors are to be involved together in the development of preventive and/or treatment programs. The focus of this program should be on active and healthy lifestyle, not directly on weight reduction, which should be viewed as a middle or long term consequence of the adoption of active habits. Perhaps, it can also be interesting to involve children themselves to the development of programs. As shown in Figure 1 the promotion of physical activity among overweight and obese youth should involve at least four important settings, including actors from each of them. An approach by the place of life seems suitable to encourage an active lifestyle among overweight and obese children. For each of these places, not only the built environment should be designed but the social climate also requires consideration so that children can freely engage in an appropriate physical activity.

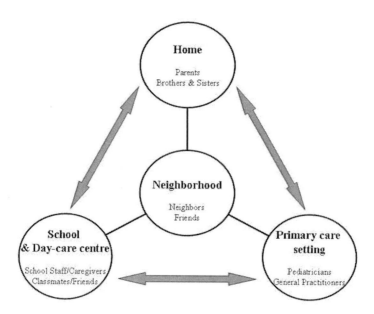

Figure 1. Main places of life and required actors for a successful active lifestyle promotion among overweight and obese children

Unlike approaches that focused solely and separately on school, family or community and whose effects are mixed, this paradigm allows putting effort together and requires a consensus from the different actors. As such, stakeholders should pursue the same objective at the same time. This requires that the political leadership is strong enough to involve the other actors whose roles are declined below, to share it and affording benefits to children.

School and Day-Care Center: Role of School Staff and Peers

Worldwide, school-age children spend the bulk of their day at school. Likewise, most preschoolers and some schoolchildren spend some times in day-care centers. The school environment as well as the day-care centre are unique in their potential to promote physical activity in children, in that they may be equipped for physical activity: play facilities, physical education classes, grounds and time for recess, and after-school opportunities for active pursuits. It also seems that it is in this environment that children spend more time in MVPA [100, 101]. All these advantages need to be adequately used in order to foster an active lifestyle among all children, and especially overweight and obese children. For this purpose, school staffs and caregivers must adopt strategies and/or policies that support or improve the physical and social environments of the school. For instance, it was found that a playground redesign using multicolor playground markings and physical structures may be a suitable stimulus for increasing children's activity levels, mainly in younger children [102-104]. Such an approach may depend mainly on the school project and the internal decision making process. On the other hand, teachers and caregivers can directly play a significant role in the promotion of an active lifestyle among overweight and obese children, by offering positive, constructive, and immediate feedback without being evaluative, critical or demanding. Correlatively, the social context of schools should be conducive so that obese children can accept themselves, and be accepted by their peers during active play times. School staffs have a crucial role to play in maintaining a good social climate between children by carefully removing all competitive spirit and avoiding stigmatization of obese children. Children should be taught how to set their personal goals that should be realistic. Overweight and obese children should also learn not to compare themselves with other children who may be more skilled. The goal needs to be self-improvement, and success should be allowed on a regular basis. For schools to keep faith, it is important to train or reinforce knowledge of teachers on the benefits of physical activity for children's health. Such knowledge should be conveyed to pupils and their parents, so that everyone can seize and support school projects designed to promote active behaviors. It is the responsibility of school policy-makers to raise the political message and encourage concrete actions such as the improvement of schoolyards, supply of equipment, and the incentive for developing authentic physical activity-related school projects.

Home: Role of Parents and the Family

The home environment can be conducive for physical activity, especially in affluent homes that are relatively large, and well-equipped for physical activity, including facilities such as yards, gardens, swimming pools, bikes, treadmills [28]. Unfortunately only a limited number of families may have these facilities in their home. Recreational facilities near apartments can be a concrete alternative for those families. Nonetheless, families should play a crucial reinforcing role in on all educational aspects, in the development and maintenance of healthy behaviors, including physical activity among overweight and obese children. Parents and the family circle of these children may directly contribute to maintaining a satisfactory level of physical activity by encouraging and regularly practicing physical activity with the child. It was found that parental physical activity had a significant influence on children's activity level. Thus children of active parents may have six times greater chance of being active themselves than their peers whose parents are inactive [105]. In addition, parental encouragement and role modeling may support the child's physical activity through their influence on the child's self-efficacy [106]. However, it should also be recognized that at a population level, the home by itself is not a promising setting for promoting physical activity [28]. Because the home is a place where sedentary attraction prevails (e.g., access to the Internet, TV or video games), children should be encouraged to go outside.

Neighborhood: Implication of Neighbors and Role of Policy Decision Makers

In the community setting, land use, proximity to destinations and neighborhood facilities may all influence the physical activity levels of children and youth. The opportunity for children to play in the neighborhood may have a significant impact on their overall level of physical activity. Neighborhood influences may include the effect of shared physical environment, the availability of services to support residents, the influence of social functioning such as social and cultural norms, and the relative effects of material deprivation [107]. The frequency of outdoors play, facilities and equipment availability, convenience of green spaces have been found to positively influence physical activity of children [108].

It should nevertheless be emphasized that an attractive and functional environment for physical activity of children can be useless due to a deleterious social climate in the neighborhood. All the community should feel concern of and be involved in the maintenance of security in the near neighborhood. To this end, the involvement of families to community programs is likely to strengthen social bonds within a given geographical area, and this social cohesion may condition the willingness of parents to encourage outdoor plays by their children

Primary Care Setting: Role of Primary Care Providers

Health professionals are key players in any community. Health professionals, and especially primary care providers and pediatricians have the responsibility of helping to prevent obesity and/or treat obese children. Primary care providers may play an integral role in working with obese children and their family, given the number of contact they have in a year with children and adolescents. In the US context, it has been reported for children <15 years, 5.7 visits/year for those with private insurance and 3.3 visits/yr for others [109]. Similar frequencies have been found in France [110, 111]. Perhaps, overweight and obese children may be even more likely to visit a primary care provider compared to non-overweight children. As such, the clinical environment should reflect a consistent health message promoting physical activity on posters and reading materials, which provide helpful tips on how to foster home environment that support physical activity [25]. Furthermore, pediatricians and other primary care providers should closely work with the obese child and his/her family and physical activity specialists - around a decisional platform represented by a municipality and its related services (e.g., schools, health service, sports service, associations). In this scheme, the role of health professionals can be particularly useful for federating other professionals around a common public health goal, which might be endorsed by the municipality and the community. The primary care provider may further translate this goal according to the child's degree of obesity, using a biopsychosocial model. The health professional should particularly increase the awareness of the child and his/her family to the risks associated with a lack of physical activity, and initiates together with them a behavioral modification approach. Such an approach requires motivation from the patient (i.e., the child and his/her parents), and it should be pointed out that a real increase of the awareness for a behavioral modification by children and their family is not sufficient. Families need to be accompanied on a regular basis. For a comprehensive health promotion plan, primary care providers could check together with families physical activity opportunities and programs that are available in local schools or the neighborhood. It is also useful and relevant that individualized counseling be provided through a collaborative decision making with family, including written prescriptions, provision of printed materials and follow-up [112, 113]. Among overweight and obese children, a behavioral change strategy is needed not only for the concerned child, but this process must also include the entire family. In this regard, the use of the 5A method [25, 114] may increase primary care professionals' efficacy toward physical activity promotion.

In summary, at an individual level, children can engage in physical activity through structured activities (*e.g.*, physical education lessons or sports) and discretionary activities (e.g., outdoor plays, playground activities). While the role of teachers and the school environment should not be neglected, parental support and role-modeling is required to reinforce the value of physical activity among children. Health professionals, and especially general practitioners and pediatricians need to learn and include physical activity counseling in their practice so as to help children and their family in incorporating an active lifestyle in their routine. Finally, it is the role of policy-makers and the entire community to create a safe and fun environment for children to enjoy outdoors activities.

Conclusion

Physical activity is a healthy behavior that all children must incorporate in their everyday life by linking it with the idea of pleasure and well-being. The balance of physical activity and other basic skills should be sought both in school/daycare and at home. But definitively, physical activity can no longer be considered secondary in an "obesogenic" context where sedentary pastimes abound. The involvement and synergy of various actors (i.e., staff at schools and daycare, neighbors, primary care providers), including children themselves is the basis for promoting sustainable active behavior in all children and in the community.

The prevention of childhood obesity and its metabolic disorders depends primarily on the acquisition and adoption of healthy lifestyles by children. Since in the short term, physical activity has a greater impact in prevention than in the correction of childhood obesity, the promotion of an active lifestyle should begin earlier during the preschool years. Taking into account the age and gender of children is a prerequisite for success since physical activity exhibit a significant decline from the early adolescence, and because girls are less active than boys. It should also be underlined that the effectiveness and dynamism of the necessary synergies for a successful physical activity program depends not only on the knowledge of the determinants and mechanisms of behavioral modification among obese children, but also on the consideration of cultural values of each context. The health promoters, without being prescriptive, should assume a coordinator and facilitator role in this respect. They should implement an evaluative process and work in a physical activity surveillance unit where behavioral changes in the pediatric population could be monitored. From an operational perspective, organizing neighborhood and school meetings can help to convey the political willpower, listen to all the actors, and finally co-produce the ingredients for promoting an active lifestyle in the community.

References

[1]   Gaziano JM. Fifth phase of the epidemiologic transition: the age of obesity and inactivity. *JAMA* 2010 303(3):275-6.

[2]   WHO. *Global health risks: Mortality and burden of disease attributable to selected major risks*. Geneva: World Health Organization; 2009.

[3]   Flegal KM, Carroll MD, Ogden CL, Curtin LR. Prevalence and trends in obesity among US adults, 1999-2008. *JAMA* 2010 303(3):235-41.

[4]   Abubakari AR, Lauder W, Agyemang C, Jones M, Kirk A, Bhopal RS. Prevalence and time trends in obesity among adult West African populations: a meta-analysis. *Obes Rev* 2008 9(4):297-311.

[5]   Ziraba AK, Fotso JC, Ochako R. Overweight and obesity in urban Africa: A problem of the rich or the poor? . *BMC Public Health* 2009 9:465.

[6]   Lobstein T, Baur L, Uauy R, IASO International Obesity Task Force. Obesity in children and young people: a crisis in public health. *Obes Rev* 2004;5 (suppl 1):4-104.

[7]   Ogden CL, Carroll MD, Curtin LR, Lamb MM, Flegal KM. Prevalence of high body mass index in US children and adolescents, 2007-2008. *JAMA* 2010;303(3):242-9.

[8]   MacPhee M. Global childhood obesity: how to curb an epidemic. *J Pediatr Nurs* 2008 23(1):1-4.

[9]   Salanave B, Peneau S, Rolland-Cachera MF, Hercberg S, Castetbon K. Stabilization of overweight prevalence in French children between 2000 and 2007. *Int J Pediatr Obes* 2009;4(2):66-72.

[10] Zimmet P, Alberti G, Kaufman F, et al. The metabolic syndrome in children and adolescents - an IDF consensus report. *Ped Diabetes* 2007;8(5):299-306.

[11] Saland JM. Update on the metabolic syndrome in children. *Curr Opin Pediatr* 2008;19:183-91.

[12] Tailor AM, Peeters PHM, Norat T, Vineis P, Romaguera D. An update on the prevalence of the metabolic syndrome in children and adolescents. *Int J Ped Obes* 2010;5(3):202-13.

[13] Weiss R, Dziura J, Burgert TS, Tamborlane WV, Taksali SE, et al. Obesity and the metabolic syndrome in children and adolescents. *N Eng J Med* 2004;350(23):2362-74.

[14] Camhi S, Katzmarzyk PT. Tracking of cardiometabolic risk factor clusrering from childhood to adulthood. *Int J Ped Obes* 2009;10:1-8.

[15] Morrison JA, Friedman LA, Wang P, Glueck CJ. Metabolic syndrome in childhood predicts adult metabolic syndrome and type 2 diabetes mellitus 25 to 30 years later. *J Pediatr* 2008;152:201-6.

[16] Oshlansky SJ, Passaro DJ, Hershow RC, Layden J, Carnes BA, Brody J, et al. A potential decline in life expectancy in the United States in the 21st Century. *N Eng J Med* 2005;352(11):1138-45.

[17] Bouchard C. Childhood obesity: are genetic differences involved? *Am J Clin Nutr* 2009 89(5):1494S-501S.

[18] Kimm SY, Glynn NW, Obarzanek E, Kriska AM, Daniels SR, Barton BA, et al. Relation between the changes in physical activity and body-mass index during adolescence: a multicentre longitudinal study. *Lancet* 2005 366(9482):301-7.

[19] Sothern MS. Obesity prevention in children: physical activity and nutrition. *Nutrition* 2004 20(7-8):704-8.

[20] Swinburn B, Egger G. The runaway weight gain train: too many accelerators, not enough brakes. *BMJ* 2004 329(7468):736-9.

[21] Chakravarthy MV, Booth FW. Eating, exercise, and "thrifty" genotypes: connecting the dots toward an evolutionary understanding of modern chronic diseases. *J Appl Physiol* 2004 96(1):3-10.

[22] Saris WHM, Blair SN, van Baak MA, Eaton SB, Davies PSW, DiPetrio L, et al. How much physical activity is enough to prevent unhealthy weight gain?Outcome of the IASO 1st stock conference and consensus statement. *Obes Rev* 2003;4:101-14.

[23] Physical Activity Guidelines Advisory Committee. *Physical Activity Guidelines Advisory Committee Report, 2008.* Washington, DC: U.S. Department of Health and Human Services; 2008.

[24] Strong WB, Malina R, Blimkie CJ, et al. Evidence based physical activity for school-age youth. *J Pediatr* 2005;146:732-7.

[25] Huang JS, Sallis JF, Patrick K. The role of primary care in promoting children's physical activity. *Br J Sports Med* 2009;43:19-21.

[26] Janssen I. Physical activity guidelines for children and youth. *Appl Physiol Nutr Metab* 2007;32:S109-S21.

[27] Majnemer A, Barr RG. Association between sleep position and early motor developement. *J Pediatr* 2006;149:623-9.

[28] Katzmarzyk PT, Baur L, Blair SN, Lambert EV, Oppert JM, Riddoch C. International conference on physical activity and obesity in children: summary statement and recommendations. *Int J Ped Obes* 2008;3:3-21.

[29] Caspersen CJ, Powell KE, Christenson GM. *Physical activity, exercise, and physical fitness: definitions and distinctions for health-related research.* Public Health Rep 1985;100(2):126-31.

[30] Corder K, Ekelund U, Steele RM, Wareham NJ, Brage S. Assessment of physical activity in youth. *J Appl Physiol* 2008 105(3):977-87.

[31] Bailey RC, Olson J, Pepper SL, Porszaz J, Barstow TL, Cooper DM. The level and tempo of children's physical activities: an observation study. *Med Sci Sports Exerc* 1995;27:1033-41.

[32] Corder K, van Sluijs EM, McMinn AM, Ekelund U, Cassidy A, Griffin SJ. Perception versus reality awareness of physical activity levels of British children. *Am J Prev Med* 2010 38(1):1-8.

[33] Reilly JJ, Penpraze V, Hislop J, Davies G, Grant S, Paton JY. Objective measurement of physical activity and sedentary behaviour: review with new data. *Arch Dis Child* 2008;93(7):614-9.

[34] Welk GJ, Corbin CB, Dale D. Measurement issues in the assessment of physical activity in children. *Res Q Exerc Sport 2000*;71(2):59-73.

[35] Trost SG. Measurement of physical activity in children and adolescents. Am *J Lifestyle Med* 2007;1(4):299-314.

[36] DuRant RH, Baranowski T, Puhl J, Rhodes T, Davis H, Greaves K, et al. Evaluation of the children's activity rating scale (SCARS) in young children. *Med Sci Sports Exerc* 1993;25(12):1415-21.

[37] Goran MI. Applications of the doubly labelled water technique for studying total energy expenditure in young children: a review. *Pediatr Exerc Sci* 1994;6:11-30.

[38] Riddoch CJ, Mattocks C, Deere K, Saunders J, Kirkby J, Tilling K, et al. Objective measurement of levels and patterns of phyiscal activit. *Arch Dis Child* 2007;92:963-9.

[39] Tudor-Locke C, Williams JE, Reis JP, Pluto D. Utility of pedometers for assessing physical activity: convergent validity. *Sports Med* 2002;32(12):795-808.

[40] Goldfield GS, Mallory R, Parker T, Cunningham T, Legg C, Lumb A, et al. Effects of open-loop feedback on physical activity and television viewing in overweight and obese children: A randomized, controlled trial. *Pediatrics* 2006;118(1):e157-e66.

[41] Adams MA, Caparosa S, Thompson S, Norman GJ. Translating physical activity recommendations for overweight adolescents to steps per day. *Am J Prev Med* 2009 37(2):137-40.

[42] Cardon G, de Bourdeaudhuij I. Comparison of pedometer and accelerometer measures of physical activity in preschool children. *Ped Exerc Sci* 2007;19:93-102.

[43] Chen KY, Bassett DR, Jr. The technology of accelerometry-based activity monitors. *Med Sci Sports Exerc* 2005;37 (suppl):S490-S500.

[44] de Vries SI, Bakker I, Hopman-Rock M, Hirasing RA, van Mechelen W. Clinimetric review of motion sensors in children and adolescents. *J Clin epidemiol* 2006;59:670-80.

[45] de Vries SI, van Hirtum HW, Bakker I, Hopman-Rock M, Hirasing RA, van Mechelen W. Validity and reproducibility of motion sensors in youth: a systematic update. *Med Sci Sports Exerc* 2009 41(4):818-27.

[46] Plasqui G, Westerterp KR. Physical activity assessment with accelerometers: an evaluation against doubly labeled water. *Obesity* (Silver Spring) 2007 15(10):2371-9.

[47] MCClain JJ, Tudor-Locke C. Objective monitoring of physical activity in children: considerations for instrument selection. *J Sci Med Sport* 2009;12(5):526-33.

[48] Guinhouya BC, Apété GK, Hubert H. Diagnsotic quality of Actigraph-based physical activity cut-offs for children: What overweight/obesity references can tell? *Pediatr Int* 2009;51(4):568-73.

[49] Guinhouya CB, Hubert H. Incoherence with studies using actigraph mti among children aged 6-12 yr. *Med Sci Sports Exerc* 2008;40(5):979-80.

[50] Guinhouya CB, Lemdani M, Vilhelm C, Durocher A, Hubert H. Actigraph-defined moderate-to-vigorous physical activity cut-off points among children: statistical and biobehavioural relevance *Acta Paediatr* 2009;98(4):708-14.

[51] Guinhouya CB, Hubert H, Soubrier S, Vilhelm C, Lemdani M, Durocher A. Moderate-to-vigorous physical activity among children: discrepancies in accelerometry-based cut-off points. *Obesity* 2006;14(5):774-7.

[52] Corder K, van Sluijs EM, Wright A, Whincup P, Wareham NJ, Ekelund U. Is it possible to assess free-living physical activity and energy expenditure in young people by self-report? *Am J Clin Nutr* 2009 89(3):862-70.

[53] Ferrari P, Friedenreich C, Mattews CE. The role of measurement error in estimating levels of physical activity. *Am J Epidemiol* 2007;166:832-40.

[54] Adamo KB, Prince SA, Tricco AC, Connor-Gorber S, Tremblay M. A comparison of indirect versus direct measures for assessing physical activity in the pediatric population: A systematic review. *Int J Ped Obes* 2009;4:2-27.

[55] Basterfield L, Adamson AJ, Parkinson KN, Maute U, Li P-X, Reilly JJ. Surveillance of physical activity in the UK is flawed: validation of the health survey for England physical activity questionnaire. *Arch Dis Child* 2008;93(12):1054-8.

[56] Ekkekakis P, Lind E. Exercise does not feel the same when you are overweight: the impact of self-selected and imposed intensity on affect and exertion. *Int J Obes* (Lond) 2006 30(4):652-60.

[57] Pender NJ, Bar-Or O, Wilk B, Mitchell S. Self-efficacy and perceived exertion of girls during exercise. *Nurs Res* 2002 51(2):86-91.

[58] Ward DS, Blimkie CJR, Bar-Or O. Rating of perceived exertion in obese and non-obese adolescents. *Med Sci Sport Exerc* 1986;18:S72.

[59] McMurray RG, Ward DS, Elder JP, Lytle LA, Strikmiller PK, Baggett CD, et al. Do overweight girls overreport physical activity? *Am j Health Behav* 2009;32(5):538-46.

[60] Ekelund U, Brage S, Froberg K, Harro M, Anderssen S, Sardinha LB, et al. TV viewing and physical activity are independently associated with metabolic risk in children: *The European Youth Heart Study*. *PLoS Med* 2006;3(12):e488.

[61] Riddoch C, Andersen L-B, Wedderkopp N, Harro M, Klasson-Heggebo L, Sardinha LB, et al. Physical activity levels and patterns of 9- and 15-yr-old European children. *Med Sci Sports Exerc* 2004;36(1): 86-92.

[62] Troiano RP, Berrigan D, Dodd KW, Masse LC, Tilert T, McDowell M. Physical activity in the US measured by accelerometer. *Med Sci Sports Exerc* 2008;40:181-8.

[63] Beets MW, Bornstein D, Beighle A, Cardinal BJ, Morgan CF. Pedometer-measured physical activity patterns of youth: a 13-country review. *Am J Prev Med* 2010 38(2):208-16.

[64] Craig CL, Cameron C, Griffiths JM, Tudor-Locke C. Descriptive epidemiology of youth pedometer-determined physical activity: CANPLAY. *Med Sci Sports Exerc* 2010

[65] Craig CL, Tudor-Locke C, Cragg S, Cameron C. Process and treatment of pedometer data collection for youth: the Canadian Physical Activity Levels among Youth study. *Med Sci Sports Exerc* 2010 42(3):430-5.

[66] Trost SG, Mciver KL, Pate RR. Conducting accelerometer-based Activity assessments in field-based research. *Med Science Sports Exerc* 2005;37(11 (suppl)):S531-S43.

[67] Janssen I, Leblanc AG. Systematic review of the health benefits of physical activity and fitness in school-aged children and youth. *Int J Behav Nutr Phys Act* 2010;7(1):40.

[68] Armstrong N, Bray S. Physical activity patterns defined by continuous heart rate monitoring. *Arch Dis Child* 1991;66:245-7.

[69] Rowlands AV, Ingledew DK, Eston RG. The effect of type of physical activity measure on the relationship between body fatness and habitual physical activity in children: a meta-analysis. *Ann Hum Biology* 2000;27(5):479-97.

[70] Dencker M, Thorsson O, Karlsson MK, Lindén C, Eiberg S, Wollmer P, et al. Daily physical activity related to body fat in children aged 8-11 years. *J Pediatr* 2006;149(1):38-42.

[71] Dencker M, Thorsson O, Karlsson MK, Lindén C, Wollmer P, Andersen LB. Daily physical activity related to aerobic fitness and body fat in an urban sample of children. *Scand J Med Sci Sports* 2008;8(6):728-35.

[72] Ness A, Leary S, Mattocks C, Blair SN, Reilly JJ, Wells J, et al. Objectively Measured Physical Activity and Fat Mass in a Large Cohort of Children. *PLoS Med* 2007;4(3):e97.

[73] Berkey CS, Rockett HR, Gillman MW, Colidtz GA. One-year changes in activity and in inactivity among 10- to 15-year-old boys and girls: relationships to change in body mass index. *Pediatrics* 2000;111:836-43.

[74] Moore LL, Gao D, Bradlee M, Cupples LA, Sundarajan-Ramamurti A, Proctor MH, et al. Does early physical activity predict body fat change throughout childhood? *Prev Med* 2003;37:10-7.

[75]   Janz KF, Kwon S, Letuchy EM, Eichenberger Gilmore JM, Burns TL, Torner JC, et al. Sustained effect of early physical activity on body fat mass in older children. *Am J Prev Med* 2009 37(1):35-40.

[76]   Houmard JA, Egan PC, Neufer PD, et al. Elevated skeletal muscle glucose transporter levels in exercise-trained middle aged men. *Am J Physiol* 1991;261:E437-E43.

[77]   Jago R, Wedderkopp N, Kristensen PL, Moller NC, Andersen L-B, Cooper AR, et al. Six-year chnage in youth physical activity and effect on fasting insulin and HOMA-IR. *Am J Prev Med* 2008;35(6):554-60.

[78]   Roberts CK, Barnard RJ, Schelk SH. Exercise-stimulated glucose transport in skeletal muscle in nitric oxide dependent. *Am J Physiol* 1997;273:E220-E5.

[79]   Shaibi GQ, Roberts CK, Goran MI. Exercise and insulin resistance in youth. *Exerc Sport Sci Rev* 2007;36(1):5-11.

[80]   Guinhouya BC. Rôle de l'activité physique dans la lutte contre le syndrome métabolique infantile. *Med/Sci* 2009;25:827-34.

[81]   Andersen LB, Harro M, Sardinha LB, Froberg K, Ekelund U, Brage S, et al. Physical activity and clustered cardiovascular risk in children: a cross-sectional study (The European Youth Heart Study). *The Lancet* 2006;368(9532):299-304.

[82]   Ebeling P, Bourey R, Koranyi L, Tuominen JA, Groop LC, et al. Mechanism of enhanced insulin sensitivity in athletes: incread blood flow, muscle glucoe transporter protein (GLUT4) concentration, and glycogen synthetase activity. *J Clin Invest* 1993;92:1623-31.

[83]   Leary SD, Ness AR, Davey-Smith G, Mattocks C, et al. Physical activity and blood pressure in childhood: findings from a population-based study. *Hypertension* 2008;51:92-8.

[84]   Mark AE, Janssen I. Dose-response relation between physical activity and blood pressure in youth. *Med Sci Sports Exerc* 2008;40(6):1007-12.

[85]   Ribeiro JC, Guerra S, Oliveira J, et al. Physycal activity and biological risk factors clustering in pediatric population. *Prev Med* 2004;39:596-601.

[86]   Nassis GP, Papantakou K, Skenderi K, Triandafillopoulou M, et al. Aerobic exercise training improves insulin sensitivity without chnages in body weight, body fat, adiponectin, and inflammatory markers in overweight and obese girls. *Metabolism* 2005;54:1472-9.

[87]   Benson AC, Torode M, Fiatarone Singh MA. Muscular strength and cardiorespiratory fitness is associated with higher insulin sensitivity in children and adolescents. *Int J Ped Obes* 2006;1:222-31.

[88]   Ahmadizad S, Haghighi AH, Hamedinia MR. Effects of resistance versus endurance training on serum adiponectin and insulin reistance index. *Eur J Endocrinol* 2007;157:625-31.

[89]   Bell LM, Watts K, Siafariskas A, et al. Exercise alone reduces insulin resistance in obese children indenpently of changes in body composition. *J Clin Endocrinol Metab* 2007 92 4230-5.

[90]   Kelishadi R, Razaghi EM, Gouya MM, Ardalan G, et al. Association of physical activity and the metabolic syndrome in children and adolescents: CASPIAN study. *Horm Res* 2007;67:46-52.

[91]   Rubin DA, McMurray RG, Harrell JS, Thorpe DE, Hackney AC. Vigorous physical activity and cytokines in adolescents. *Eur J Appl Physiol* 2008;103:495-500.

[92] Brage S, Wedderkopp N, Ekelund U, et al. Features of the metabolic syndrome are associated with objectively measured physical activity and fitness in children: the European Youth Heart Study (EYHS). *Diabetes Care* 2004;27:2141-8.

[93] Ekelund U, Anderssen S, Andersen L-B, Riddoch C, et al. Prevalence and correlates of the metabolic syndrome in a population-based sample of European youth. *Am J Clin Nutr* 2009;89:90-6.

[94] Ekelund U, Anderssen SA, Froberg K, Sardinha LB, Andersen L-B, et al. Independent associations of physical activity and cardiorespiratory fitness with metabolic risk factors in children: the European youth heart study. *Diabetologia* 2007;50(9):1832-40.

[95] Roemmich JN, Barkley JE, Lobarinas CL, Foster JH, White TM, Epstein LH. Association of liking and reinforcing value with children's physical activity. *Physiol Behav* 2008;93:1011-8.

[96] Benezech J-P. L'activité physique diminue la douleur. *Douleurs* 2005;6(3):141-4.

[97] Venojarvi M, Puhke R, Hamalainen H, Marniwmi J, et al. Role of skeletal muscle-fiber type in regulation of glucose metabolism in middle-aged subjects with impaired glucose tolerance during long-term exercise and dietary intervention. *Diab Obes Metab* 2005;7:745-54.

[98] Goldfield GS, Mallory R, Parker T, Cunningham T, Legg C, Lumb A, et al. Effects of modifying physical activity and sedentary behavior on psychosocial adjustment in overweight/obese children. *J Pediatr Psychol* 2007 32(7):783-93.

[99] Benson AC, Torode M, Fiatarone Singh MA. Effects of resistance training on metabolic fitness in children and adolescents: a systematic review. *Obes Rev* 2008;9:43-66.

[100] Fein AJ, Plotnikoff RC, Wild C, Spence JC. Percieved environment and physical activity in Youth. *Int J Behav Med* 2004;11(3):135-42.

[101] Guinhouya BC, Lemdani M, Apété GK, Durocher A, Vilhelm C, Hubert H. How school time physical activity is the "big one" for daily activity among schoolchidren? A semi-experimental approach. *J Phys Activity Health* 2009;6(4):510-9.

[102] Ridgers ND, Stratton G, Fairclough SJ, Twisk JW. Long-term effects of a playground markings and physical structures on children's recess physical activity levels. *Prev Med* 2007;44(5):393-7.

[103] Ridgers ND, Stratton G, Fairclough SJ, Twisk JW. Children's physical activity levels during school recess: a quasi-experimental intervention study. *Int J behav Nutr Phys Act* 2007;4:1-9.

[104] Stratton G, Mullan E. The effect of multicolor playground markings on children's physical activity level during recess. *Prev Med* 2005;41(5-6):828-33.

[105] Moore LL, Lombardi DA, White MJ, et al. Influence of parent's physical activity levels on activity levels of young children. *J Pediatr* 1991;118:215-9.

[106] Trost SG, Sallis JF, Pate RR, Freedson PS, Taylor WC, Dowda M. Evaluating a model of parental influence on youth physical activity. *Am J Prev Med* 2003;25(4):277-82.

[107] Flowerdew R, Manley DJ, Sabel CE. Neighborhood effects on health: does it matter where you draw the boudaries? *Soc Sci Med* 2008;66:1241-55.

[108] Sallis JF. Epidemiology of physical activity and fitness in children and adolescents. *Crit Rev Food Sci Nutr* 1993;33(4-5):403-8.

[109] Schappert SM, Burt CW. Ambulatory care visits to physician offices, hospital outpatient departments, and emergency departments: United States 2001-02. *Vital Health Stat* 2006;13(159):1-66.

[110] Debout C, Omalek L. Plus de recours aux médecins spécialistes en Ile-de-France. *INSEE Ile-de-France à la page* 2005;247 1-4.

[111] Galli J, Debout C, Omalek L. Plus de recours aux médecins spécialistes en Ile-de-France. *Insee Ile-de-France, Dossiers* 2007:44-7.

[112] Jacobson DM, Strohecker L, Compton MT, Katz DL. Physical activity counseling in the adult primary care setting: position statement of the American College of Preventive Medicine. *Am J Prev Med* 2005 29(2):158-62.

[113] Whitlock EP, Orleans CT, Pender N, Allan J. Evaluating primary care behavioral counseling interventions: an evidence-based approach. *Am J Prev Med* 2002 22(4):267-84.

[114] Guinhouya BC. Physical activity counseling in the primary care setting: Insight into the 5A's method. *Educ Ther Patient/Ther Patient Educ* 2010; 2010;2(2):S201-S11.

In: Childhood Obesity: Risk Factors, Health Effects…   ISBN: 978-1-61761-982-3
Editor: Carol M. Segel   ©2011 Nova Science Publishers, Inc.

Chapter II

Contribution of Physiological and Psychosocial Risk Factors Combined With Sedentary Lifestyles to Overweight and Obesity: Prospects for Prevention

Pouran D. Faghri and Cynthia H. Adams
University of Connecticut, Storrs, Connecticut, USA

Purpose: To examine the co-morbidity of the physiological risks tied with psychosocial and environmental factors and present a prospective for prevention of overweight, obesity and related disorders.

Objectives:

1. Describe obesity as a complex phenomenon with multiple and interacting risk factors.
2. Describe the significant increase in sedentary lifestyle coupled with high consumption and availability of calorie dense foods as the catalyst to an obesity epidemic in US.
3. Demonstrate that an understanding of weight gain and weight management with their complexities and multiplier effects is imperative.
4. Review the genetic base for understanding obesity which has advanced in the past decade due to the scientific discovery of several genes that regulate body weight.
5. Demonstrate the environmental contribution to obesity which has also been advanced as society promotes increased calorie consumption and decreased physical activity.
6. Understand the psychosocial factors, such as forced conformity to unrealistic norms, denial of success and access to those who do not fit a standard; and the association of mood, serotonin, and food as medicine to suppress depression which must also be factored in when proposing a novel approach for the prevention of obesity.

7. Demonstrate use of the multiplier approach, where genetics, environmental and psychosocial factors are addressed.
8. Recommend strategies to address the obesity epidemic in US.

Introduction

Over the past 25 years, the prevalence of obesity has been increasing at an alarming rate (Ogden et al., 2007). The number of adults who are obese is dangerously high, with approximately one-third of adults catogized as obese in 2007-2008 (Flegal, 2010) and an even more dire concern arises over the nearly 20% of children ages 6 – 11 years and 18% of teenagers who are now clinically obese (Childhood Obesity Statistics, 2010; Childhood Overweight and Obesity, 2010).

Excess body fat is linked to several health risks including hypercholesterolemia (high blood cholesterol), hypertension (high blood pressure), type 2 diabetes, coronary heart disease, and stroke. As the rate of obesity escalates the prevalence of these related health indices is also on the rise (CDC, 2009). To further elucidate the impact of weight gain on US citizens the statistics on weight change available for the last 25 years and patterns of DMT2 (Diabetes Mellitus Type2) are precisely parallel occurrences (Klein, et al., 2004; The Writing Group, 2007; Ogden, et al., 2008). Furthermore, in the United States, obesity has been estimated to cost as much as $147 billion per year (Finkelstein, Trogdon, Cohen, & Dietz, 2009). Government and private insurers are left to finance the economic burden associated with obesity. Therefore, government, taxpayers, employers and employees alike, are invested and supportive of obesity prevention. A variety of integrated factors play a dynamic role in obesity, making it a complex health issue to address. This chapter will examine the co-morbidity of the physiological risks tied with psychosocial and environmental factors and present a prospective for prevention of overweight, obesity and related disorders.

Physical Activity and Obesity

Inadequate physical activity has become the norm among all age groups. Daily physical activity requirements were once being met at the workplace and home with society's dependence on ambulatory transportation. However, today, the demand for physical activity has nearly ceased. Energy expenditure in the workplace is minimal due to a greater preference for maximizing the time and productivity of employees. Development of new technologies including computers and labor-saving devices has also increased the tendency for a more sedentary work environment. In addition to the decrease in energy expenditure at the workplace, leisure-time physical activity has also declined. In the United States, only 4 out of 10 adults meet physical activity guidelines. Electronic entertainment systems (i.e. televisions, computer games, and the Internet) have contributed significantly to this trend toward a more sedentary lifestyle. It is estimated that children aged 8–18 y spend >3 h per day on average watching TV, DVDs, and movies and playing video games (Roberts, Foehr, & Rideout, 2005). Children spend the majority of their day in school; however, the percentage of high school students who attended physical education classes daily decreased from 42% in 1991 to

25% in 1995, and has remained stable at that level until 2005 (33%). In 2005, while 45% of 9th grade students attended physical education class daily, only 22% of 12th grade students did so. Of students who attended physical education classes, 84% actually exercised or played sports for 20 minutes or longer during an average class (CDC, 2006). Consequently, the shortfall in physical activity is considered a noteworthy factor associated with the rising obesity epidemic.

To accurately deduce that the rising obesity trend is related to a decline in physical activity it is necessary to establish that a definite relationship exists between positive energy expenditure and reduction in obesity. Studies have been conducted which validate the previous statement; thus, confirming the inverse relationship between physical activity and obesity trends. Lakdawalla and Philipson (2007) suggested that after 18 years, an average male worker would weigh 25 pounds more working in the lowest fitness-demanding jobs rather than working in the most physically demanding jobs. The Australian Diabetes, Obesity and Lifestyle Study concluded that, on average, a 10% increase in sedentary time was associated with a 3.1 cm (95% CI 1.2-5.1) increase in waist circumference (Healy et al., 2008). Overall, new technologies enhance productivity at the expense of employees' health.

A recent study by Kwon and colleagues (2010) reported that daily moderate intensity aerobic exercise effectively reduced abdominal adiposity in obese diabetic patients. Furthermore, Jackicic et al. (2010) examined the effect of different durations of moderate-to-vigorous intensity physical activity on body weight in overweight adults (BMI = 25.0 to <30.0 kg/m2), independent of diet. The results of this study did not find significant weight loss differences among the groups; however, investigators were able to conclude that 150min/week or 300min/week of physical activity resulted in a greater than 2% weight loss among participants. These findings were also consistent with an earlier literature review conducted for the Physical Activity Guidelines for America, stating that ≥ 180 minutes/week of physical activity typically resulted in 1-3% loss of initial body weight. In a secondary analysis participants were categorized based on whether they maintained ($\pm 3\%$ of baseline weight), gained (>3% of baseline weight) or lost (<3% of baseline weight) (Jackicic et al., 2010) weight. At the18-month follow-up, weight change was $0.0 \pm 1.3\%$, $+5.4 \pm 2.6\%$, and $-7.4 \pm 3.6\%$, respectively while change in physical activity was 78.2 ± 162.6 min/week, 74.7 ± 274.3 and 161.9 ± 252.6 min/week. Jackicic and colleagues (2010) suggested that compliance with increased physical activity (161-228 minutes/week above baseline) significantly contributes to weight loss. Although, this study did not impose any dietary restrictions, those individuals who lost weight also showed improved scores on the Eating Behavior Inventory. This outcome implied that individuals engaging in a physical activity regimen may be more inclined to make healthy dietary changes or that exercise decreases intake. This notion is consistent with the idea posed by Baker and Brownell (2000) that increases in exercise self-efficacy will transcend to other weight loss behaviors such as eating. It is also important to note that these changes in eating patterns in combination with increased physical activity are likely to have an additive effect on promoting weight loss. Overall, significant evidence confirms the inverse relationship between physical activity and obesity and supports the parallel relationship between obesity trends and the shift to sedentary lifestyle.

Environmental Contributions to Obesity

Today's environment is largely conducive to behaviors that are obesity promoting. Obesity experts describe our environment as obesogenic (Brownell, 2004; Booth, 2001). This environment is both economically and biologically driven. There is widespread availability of unhealthy and affordable high-energy foods and limited access to parks, sidewalks, and recreational facilities which support physical activity. Consequently, these environmental factors offset the energy balance and have contributed to the growing obesity epidemic. Studies have shown that better access to supermarkets is related to a decreased risk of obesity in comparison to closer proximity to convenience stores which is associated with a greater risk of obesity (Morland, Diez Roux & Wing, 2006; Powell, Auld, Chaloupka, O'Malley, & Johnston, 2007). Morton and Blanchard described lower income, minority urban and rural neighborhoods as "food deserts" due to the lack of conveniently located supermarkets. Marcellini et al. (2009) observed a relationship between socioeconomic status and obesity risk. Statistical analysis concluded that obese subjects had a higher rate of low-socioeconomic status in comparison to non-obese subjects. Brownell (2004) stated that the higher proportion of fast-food restaurants may be supporting the obesity epidemic; with a higher number of fast-food restaurants located in poorer neighborhoods (Powell, Chaloupka & Bao, 2007). Overall, accessibility, especially to high-calorie, high-fat food options offered at convenient stores and fast-food restaurants, may explain why there is a difference in the degree of obesity between these divergent economic communities. Franco et al. (2009) concluded that there was an association between corner-store offerings with poor overall dietary intake. In a study published in 2010 by Lucan et al., a detailed inventory was taken and a nutrient analysis was performed of the snack foods available at seventeen corner-stores in low income minority neighborhoods in Philadelphia. Focus groups were held with local students to determine the corner-store they most commonly patronized and the food items they typically purchased. The researchers reported that no fruit/vegetable snacks were available at any of the corner stores, only 3.6% of the snack food contained whole grain while the remaining snacks (96.4%) were processed foods. Nutrient analysis of the snack options showed that consuming one snack item could provide half a day's caloric intake, and approximately 100% of the recommended limits of fat, sugar, and sodium (Lucan et al., 2010). Furthermore, in an earlier study, Borradaile and colleagues (2009) identified that children, on average, purchase 1.6 food items per sale, nearly every day and 42% make two purchases per day. Based on the findings from these two studies it can be observed that a significant portion of calories consumed by children are based on poor snack options which are readily accessible and may be their only options.

Another factor driving poor food choices and procurement is cost. Since the late 1980s, in the United States, the price of fresh fruit and vegetables has increased by 190%, all fruit and vegetables by 144%, fish by 100%, and dairy products by 82%. In comparison, the price of fats and oils only increased by 70%, sugars and sweets by 66%, and carbonated beverages by 32% (Finkelstein & Zuckerman, 2008). Consumers tend to make food choices based more on economic costs and taste reward, rather than on health. Currently, there has been a shift to reliance on cheaper, calorie-dense foods, rather than healthful alternatives (Popkin, 2008). The Centers for Disease Control and Prevention reported that between the late 1970s and today, men, women, boys and girls have all increased their daily food intake by 80, 360, 250, and 120 calories, respectively. This increase in caloric intake can be attributed to the greater availability of high energy-dense foods. Mokdad et al (2001) found adults self-reported that only 24% consumed 5 or more fruits and vegetables per day. Without a positive change in energy expenditure to compensate for these additional calories, weight gain is inevitable.

Overall, environmental factors such as a reduced need for physical activity at the workplace and at home as well as limited access to sidewalks, walking/biking trails, parks and recreational facilities have reduced the amount of energy individuals expend. Additionally, the access, availability and affordability of high-energy, low nutrient food increases the amount of calories consumed. The interplay of all these environmental factors leads to a positive energy balance and an increased risk of obesity. An individual's genetic background is thought to play a dynamic and interrelated role in obesity risk. Data suggests that 40-70% of BMI variation is genetically dependent, while at least 30% is due to cultural and environmental factors (Comuzzie & Allison, 1998; Hill & Peters, 1998). The following section will discuss the pivotal role genetics has on obesity risk.

Genetics and Obesity

New research has been able to identify numerous genes directly and indirectly related to obesity. Genes related to obesity risk are responsible for energy metabolism, appetite, taste, metabolic and signal pathways, and adipogensis. There is evidence that these genetic variations make some individuals more susceptible to weight gain than others.

Recently, a genome-wide study identified the fat mass and obesity gene (FTO). This gene has been the strongest common genetic predictor of obesity known so far (Frayling, Timpson, Weedon, et al., 2007). The FTO gene is dominant in the hypothalamus and has been shown to be associated with appetite. In a cross sectional study, Sonestedt et al. (2009) examined whether macronutrient intake and leisure-time activity alter the genetic variation in FTO and obesity. Findings from this study signified the relationship between fat intake and genotype on BMI. In addition, this study suggested that a low carbohydrate diet was associated with higher BMI whereas a higher carbohydrate diet had no significant affect on BMI. Also, based on self-reported physical activity, investigators confirmed earlier study findings linking FTO genotype and increased BMI with low physical activity. Accordingly, the FTO gene may be manipulated by dietary macronutrients and exercise.

This relationship between genes and lifestyle interactions and their influence on obesity risk has also been studied in individuals carrying the Trp64Arg polymorphism in the beta-3-adrenoceptor gene and the Pro12Ala mutation of the PPARgamma gene or the Gln27Glu polymorphism for the Beta-2-adrenoceptor gene. Individuals carrying the Trp64Arg polymorphism in the beta-3-adrenoceptor are at significant risk of obesity when they remain sedentary (Marti, Corbalan, Martinez-Gonzalez & Martinez, 2002). High carbohydrate diets increase obesity in individuals with the Pro12Ala mutation of the PPARgamma gene or the Gln27Glu polymorphism for the Beta-2-adrenoceptor gene (Marti, Corbalan, Martinez-Gonzalez, Forgal & Martinez, 2002; Martinez, Corbalan, Sanchez-Villegas, Forga, Marti, & Martinez-Gonzalez and, 2003).

In a more recent study Tanofsky-Kraff et al. (2009) examined the phenomenon of loss of control (LOC) eating in relationship with the FTO rs9939609 obesity risk allele. The investigators found no significant difference in overall calorie consumption between those who did and those who did not possess the obesity risk allele. However, consistent with previous findings (Sonestedt et al., 2009), fat intake was higher in individuals with the FTO gene. Additionally, this study concluded that children with at least one of the obesity risk alleles were more likely to experience LOC eating. Tanofsky-Kraff et al. (2009) postulated that decreasing LOC eating may be effective for body weight loss and obesity prevention.

Current environmental and lifestyle changes are suggested to influence the function of the "stress" genes which further contribute to the current epidemic of obesity (MacEwen, 2000). One form of stress is directly related to how many hours American's sleep. The reported number of hours slept has progressively decreased (Cizza, Skarulis & Mignot, 2005). Inadequate sleep influences the release of the stress hormone cortisol increasing central adiposity, while decreasing leptin and increasing ghrelin, stimulating appetite and food intake, respectively (MacEwen BS, 2006). In turn, the increased availability of food, decrease in physical activity, and inadequacy of sleep have resulted in a state of positive energy balance.

As previously discussed there is wide accessibility to high-calorie foods. The most influential determinant manipulating food intake is taste (Glanz, 1998). From birth, there is an innate satisfaction associated with sweetness and a displeasure related to bitterness (Steiner, 2001). Ultimately, people consume foods which they find most pleasant and what they select to eat determines their nutritional status and their risk of obesity and chronic diseases; therefore, certain food preferences result in greater health implications such as obesity.

Currently, research is being conducted which proposes that genetics plays a role in predicting individuals' food preferences. Sensitivity to bitter tasting compounds, such as phenylthiocarbamide (PTC) or 6-n-propylthiouracil (PROP) is a well recognized genetic trait. The different perceptions of bitterness were found to result from a genetic polymorphism of the *TAS2R38* receptor. Duffy et al. (2004) determined that PROP bitterness varied significantly across genotypes; thus, the AA homogenous taste less bitterness than either AP heterogeneous or PP homogenous. Humans also have unique variations in the number of fungiform papillae and taste buds within these papillae (Duffy, 2007). This variation plays a role in the different oral sensory abilities experienced by individuals. Super-tasters have a greater number of these papillae and taste concentrated PROP as intensely bitter, while non-tasters taste it as weakly bitter and have fewer fungiform papillae (Duffy, 2007). As a result, individuals naturally find certain foods more palatable. Fischer et al. (1966) noted that PTC tasters tended to be ectomorphs (thin and angular body types), whereas non-tasters tended to be endomorphs (generous body proportions). Additional studies suggest that greater weight among PROP non-tasters could result from higher preferences and intakes of sweets, alcohols and alcoholic beverages (Duffy, 2004).

Psychosocial Factors and Obesity

The psychological effects of obesity in children are devastating and one of the most socially stigmatizing conditions of childhood (Schimmer, et al., 2003). Depression and childhood obesity are correlated and obesity and psychiatric disorders lead to Oppositional Defiant Disorder (Mustillo, 2003). Weight and shape are very important to the value adolescents hold of themselves, with those reporting binge eating (also referred to as LOC eating) having significantly lower scores on measures of body satisfaction and self-esteem and higher scores on measures of depressive mood. Further, overeating was associated with suicide risk in a study by Ackard and colleagues (2003). No child wishes to be obese and the daily depression and low self-esteem which accompany childhood obesity is sufficient reason alone for medical intervention even if diabetes and other disorders were not tied to the overweight.

In 1980: 2 – 5 year olds were diagnosable as obese in only 5% of cases; 6 – 11 year olds at 6.5%; and teenagers also at 5%. However, sixteen to eighteen years later 2 – 5 year olds were 12.4% obese; 6 -11 year olds were 19.6 % obese; and teenagers were 18% obese (Childhood Obesity Statistics, 2010; Childhood Overweight and Obesity, 2010). Childhood obesity has led to a 3 fold increase in medical costs for children since 1990 and a rise in DMT2 in children under age 10 thus warranting a serious concern for treatment (Mayer-Davis, 2001).

It must be noted that when children develop DMT2 the consequences are dire. The pediatric patient with DMT2 experiences the consequences of diabetes for a more prolonged period than do typical Type 2's and a more rapid advance of the complications of diabetes resulting in cardiovascular health issues previously expected in geriatric populations (Klein, et. al., 2004; Song & Hardisty, 2008).

Binge eating disorder (BED) is positively related to obesity and to psychological issues. In a study by Colles, Dixon and O'Brien (2008) of LOC eating, participants with severe emotional disturbance due to feelings of LOC reported higher symptoms of depression and poorer mental health. Psychological distress and self-condemnation along with elevated symptoms of depression and higher body-image concerns were commonly associated with BED (Antony, et al., 1994; Mussell, et al., 1996; Eldredge & Agras, 1998; Wilfley, et al., 2000; Specker, et al., 2004; Colles, Dixon & O'Brien, 2008).

Fairburn and colleagues published a study in 1998 using a case-control design of 52 women with BED, compared with 104 women without an eating disorder, 102 women with other psychiatric disorders and 102 women with bulimia nervosa. When BED women were examined against the healthy controls they reported a higher incidence of adverse childhood experiences, parental depression, and repeated exposure to negativity about their weight and eating and they were vulnerable to obesity. Compared to the women with other psychiatric issues the BED women were higher on more exposure to negative comments about body image and had already experienced childhood obesity. The pronounced vulnerability to obesity was the most significant distinction between the BED women and those with traditional bulimia nervosa (Fairburn, 1998).

Emotional eating is a psychological problem and the effects of this lead to obesity. It is well accepted that the consumption of particular foods stimulates the brain to release extra serotonin. Serotonin is a neurotransmitter: Too little serotonin has been linked to aggressive behavior and depression; or if serotonin receptors are not processing serotonin effectively this will also cause depression (Young, 2008). Serotonin can be increased by eating chocolate, drinking milk or eating nuts and whole grains. This is part of the psychological reward of eating that alters and improves mood. Endorphins are also natural mood enhancers released by the brain and occur naturally when the body is combating pain. Three healthy ways to release endorphins are through physical activity to the level of perspiring; eating spicy foods; and laughter for an extended period of time (Rodriguez-Baziuk, 2009). Forced vomiting may also lead to an endorphin response and is one way that those suffering from Bulimia Nervosa may become addicted to the purge phase.

The obese score higher on a multitude of factors indicating poor emotional health (Butler, Vallis & Perey, 1999); the quality of their lives is affected by social condemnation and by physiological pain related to the stress on their bodies (Barofsky, Fontaine, & Cheskin, 1998); and the health related quality of their lives is a serious condition from which chronic illnesses emerge (Larsson, Karlsson & Sullivan, 2002). The lower quality of life and the concomitant depression associated with obesity (Wadden & Phelan, 2002) make it essential to continue to examine and evaluate remedies for this chronic condition.

Multifactoral Strategies for Change

Risk factors must be viewed in four major categories: Biological, genetic, environmental, and psychosocial. The national epidemic of obesity and metabolic syndrome require a multifactoral approach for prevention and/or remediation.

Biology: Biologically, the human body is designed to maintain energy balance in the context of both the body's energy conservation and daily physical activities. The human body is also developed to promote food intake and food restriction is not part of its genetic/biological mission. Humans have strong preferences for fat and sweet and foods with high energy to fight starvation. Biologically, the human body is not built to be active unless to access food or to defend itself (Hill, 2002; Tarasuk, 1991; Drewnowski, 1998).

Genetic: While heredity may dictate how food tastes, what is preferred, a basic body type, and to some extent metabolism itself; education and self-determination are capable of modifying heredity (Sonestedt, 2009). That is, a person with overweight family members and a history of diabetes and coronary artery disease may make the decision to choose foods wisely, get regular exercise, consume preventive medications and lead an active life and: They will need to know what effects their genes have on metabolism to know which macronutrients to emphasize. Their struggle may be continuous but the potential rewards are lifesaving. Thus far there are no routine ways to alter one's genetic composition.

Recognition of the potential effect of heredity is a key to taking the proactive steps necessary for prevention. If this is to be effective, parents must be engaged in positive health behaviors in order to teach their children before obesity has occurred; or to alter this condition while their child is young. Recall individuals carrying the Trp64Arg polymorphism in the beta-3-adrenoceptor are at significant risk of obesity when they remain sedentary (Marti, Corbalan, Martinez-Gonzalez & Martinez, 2002). Vacations and weekends focusing on outdoor activities, athletics, and active community events get children up and away from the television and computer games. Parents of young children need support, education and resources to employ this strategy (Ponder & Anderson, 2008; Risks, 2010). In all probability the parents too are overweight and endeavors to aid their child will also have a positive effect on all family members. While a detailed discussion of genes may be counterproductive with the lay public, health professionals need to synthesize this information to be of assistance.

Environmental: Overweight families may not only posses genetic tendencies toward weight but may also tend to serve larger portions of food than do thinner cohorts. When this was an agrarian society with families involved in intense labor on the farm a quarter slice of a pie was minimal before work began: But when workers left manual labor to sit at desks and computers the family may have continued to offer the same amount of food. The environment and culture did not shift their norms with the reduced energy needs.

This is also the age of "Super-Sized" meals (Young & Nestle, 2002). As stated earlier the all too easy access to high calorie dense foods and the constant promotion of these items makes it challenging to avoid them. Foods that had once been low or no calories such as a cup of coffee have also been altered and taste enhanced so that they are constantly tempting and fattening.

When the environment entices individuals to consume more calories than are needed and also offers decreased need for energy expenditure, as described earlier in this paper, society is faced with an 'obesogenic' environment (Hetherington, 2007). The amount eaten depends less on hunger and need and more on environmental context and availability. Environmental changes need to focus on physical activity and accessibility of fresh fruits and vegetables and less on empty calories where weight is gained and no nutritional value is.

Environmental factors can play a significant role is increasing physical activity. Communities within close proximity of parks, recreation facilities and designed access to sidewalks have been touted for promoting health enhancing behaviors. These communities have been associated with higher levels of physical activity and in turn, lower risks of overweight and obesity (Frank, 2004; Gordon-Larsen et al., 2006). Moreover, Frank and colleagues (2005) reported that among adults living in high versus low *walkability* neighborhoods, 37% versus 18%, respectively, were meeting physical activity recommendations. In another study, Handy, Cao and Mokhtarian (2008) explored the relationship between environment and behavioral changes. By following individuals who moved to more *walkable* neighborhoods, they were able to report an improvement in physical activity levels whereas individuals who moved to less walkable neighborhoods experienced a decline in physical activity. Handy, Cao and Mokhtarian (2008) were able to show that environmental factors which are conducive to physical activity will in fact stimulate a positive behavioral change. Accordingly, creating walking paths/trails, sidewalks and access to parks and recreational facilities can motivate behavioral change. Reducing perceived barriers which prevent physical activity may increase the amount of physical activity and offset the disproportionate energy balance.

In a recent study Annesi and Unruh (2007) tested a model presented by Baker and Brownell which hypothesized relationships between physical activity, alteration in psychological factors (i.e. positive changes in mood, body-image and exercise-related self-efficacy) and weight-loss. In this study the treatment group was provided with cognitive behavioral treatment (ie. contracting, stimulus control, cognitive restructuring and dissociation from exercise induced discomfort), nutrition information sessions, exercise planning and access to a wellness center. Findings were in accordance with Baker and Brownell's model. Positive changes in mood, body image and self-efficacy were correlated with exercise attendance. The hypothesized relationship between changes in depression scores with weight and body composition changes were partially supported. This is important to note because some evidence shows that depression and an overall low mood are associated with high-fat and high-calorie diets. Psychological distress and low mood tends to stimulate consumption of high-calorie and high-fat foods as a comfort mechanism.

Psychosocial: The development of weight management programs has been extensively studied; yet, most have failed to show prolonged weight loss maintenance. Some commercially successful programs seem rather to profit from repeat customers (Orbach, 2002). Perri and colleagues attempted various treatment options including extended treatment contact, social support, and exercise and training in "relapse prevention," whereas other researchers tried extended contact with a therapist, monetary incentives, controlled energy intake, telephone contacts and the use of personal trainers to encourage exercise (Baum, Clark & Sanders, 1991).

Research has been conducted regarding psychological and behavioral characteristics predictive of success and failure with weight-loss. Today's society promotes a preoccupation with obtaining unrealistically low weights. According to the National Institute of Health (NIH), losing 5-10% of body weight can significantly improve cholesterol, blood pressure, blood glucose and other health problems related to obesity and further evidence shows that these benefits are sustained with weight loss maintenance (NIH 2007). Yet most individuals enter weight loss programs with the desire to lose 20-30% of their initial body weight (Jeffery, Wings & Mayer, 1998).

Sufficient evidence suggests that being overweight and obese increases health risks; however, few females express "improving health" as a motive for losing weight. Cooper and Fairburn (2001) found that women, in particular, seek obesity treatment primarily to change their appearance. This drive to be physically content with one's shape is deemed necessary in order to achieve other goals. Moreover, the desire to achieve unobtainable weight-loss goals may lead to the abandonment of all weight-loss efforts. Studies have shown that dichotomous thinking, perceived barriers, negative self-evaluation of worth based on body image, and emotional eating were associated with weight loss failure (Byrne, Cooper, & Fairburn, 2004; Kruger, Blanck, & Gillespie, 2006); whereas, self-efficacy and self-management strategies were related to weight loss success (Tiexiera, Going, Hout, Cussler, Martin, & Metcalfe, 2002). For some, the inability to meet weight loss goals is considered a total failure. Rather than acknowledging and rewarding small accomplishments they regard their efforts as futile and anything less than their goal as worthless leading to abandonment of all efforts.

A prospective cohort study by Byrne and colleagues (2004) investigated fifty-four women with obesity that had lost 10% of their initial body weight while participating in a community weight loss program. Findings from this study showed dichotomous thinking was the strongest predictor of weight regain with the most significant association in those participants who experienced a great degree of dichotomous thinking at the time of at least 10% weight loss. Byrne postulated that individuals with dichotomous thinking styles may predict their own abandonment of weight loss efforts. Dichotomous thinkers have a tendency to perceive their short comings as total failures and to consider their achievements as inadequate. In this study, prior to weight regain, participants were less satisfied with their weight and demonstrated a lack of weight control compared with those who maintained their weight. Additionally, a higher maximum weight was also significantly linked with weight regain. The results of the study by Byrne et al. support the notion that psychological factors influence the individual's inability to maintain weight loss.

Difficulty in maintaining weight loss suggests at least one strategy for improved health must focus on prevention with children. The concept of limiting adipose cites before they develop is efficacious however recent pediatric obesity statistics reveal a problem that has already spiraled as reviewed earlier. Studies are attempting to define diet modifications which make foods, especially fruit and vegetable, more palatable to those who are classified as non-tasters, as a way of lowering their dietary risk of obesity and cardiovascular disease.

Satiety is easily interrupted by offering a variety of foods, having others present or introducing tasks which must be accomplished during the meal. Overconsumption is enhanced by an abundance of highly-palatable, energy-dense foods, alcohol use and cues that stimulate the individual to binge (Hetherington, 2007; Colles, Dixon, & O'Brien, 2008). Strategies for decreasing consumption are not impossible to develop but require that the subject be highly motivated to change.

Overeaters Anonymous works as do many support groups: It offers social support for people attempting to recover from a substance abuse, in this case food. Learning to modify their diets as mentioned briefly above must become a way of life and not a temporary change. Social gatherings will always feature food and large displays are a likely trigger for the obese to overeat but with support the overweight can learn to make the right choices and buffets that offer grilled fish, chicken, fruits and vegetables make it possible for all to eat in a healthy manner. Part of what will make this work is a shift in societal norms so that all family members are health conscious and activity, rather than food, focused; families that eat together and with the parents modeling good eating for their children (Risk, 2010).

Strategic Framework for Intervention: Obesity is attributed to a multitude of factors; therefore, the first step in decreasing the rate of obesity is to understand the relationship between these variables and work in collaboration to address the issues as described above. Presently, most of our interventions are at individual levels without concurrently addressing ubiquitous economic, physical, environmental, food industry, cultural and food policies. Most obesity interventions have been encouraging the person to resist the obesogenic environment at work, home and during leisure times (Brownell, 2004; Booth, 2001). These interventions provide resources on a short-term bases that are proven effective; however as soon as the resources are expended the individual returns to their old behaviors and will regain weight. Addressing the obesity epidemic is multifactoral and requires individual and community/governmental policy interventions to be effective. The following list of strategies addresses the obesity epidemic:

[1] Reduce positive energy balance by incorporating proven behavioral theories to increase individual's level of information and awareness and motivate them to change their behavior to combat positive energy balance.

[2] Provide an environment that encourages physical activity such as sidewalks, walking paths with proper lighting and discourage sedentary lifestyle.

[3] Use behavioral economics toward reducing obesity. Encourage society to say "no" to large portions of food for less weight, and a better quality of life. Saving money but adding weight is a poor investment.

[4] Provide an environment at work and school that encourages healthy weight (i.e. banning soda from schools and workplace).

[5] Work with food and the grocery industry to promote healthy food choices. Reward healthy shopping with extra cash or more healthy choices (for example, choosing five fruits and vegetables in your grocery, will give you money back in your next purchase). Ban or limit advertisement for unhealthy foods in these settings. Do not market "junk" foods to children.

[6] Use incentives to promote healthy behavior in the workplace. Develop policies that provide incentives for maintaining and engaging healthy lifestyle behavior. It is the norm to receive incentives for being productive and adhering to policies. If the policy

of the workplace is healthy behavior why not reward healthy behavior the same way as productivity?

[7] Encourage healthy behavior at school. Behaviors begin and become entrenched during childhood. Schools could be a key environment to address obesity epidemics. Studies have shown that better lifestyles (both diet and exercise) promote academic success. This could be used as an incentive to collaborate with families and schools to provide healthy lifestyle education and interventions for children.

[8] Encourage healthy behavior at the community level. Communities have a major impact on people's lives. Work with communities to provide infrastructures that promote and build environments for active lifestyles, and easy access to local and healthy foods (such as farmers' markets).

[9] Work with local, state and federal governmental agencies to provide provisions and policies that promote and encourage society to engage in lifestyle activities that reduce obesity. Use smoking policies and provisions as examples.

Summary and Final Thoughts

Biological, behavioral, and environmental factors all contribute to the complexity of alleviating the obesity epidemic. Society provides an environment that encourages over-eating and sedentary lifestyle. Despite the recognition of the seriousness of obesity to public health, no interventions have been effective in reducing obesity rates on a population basis. The obesogenic environment is motivated both economically and biologically. Biological preferences are built into the environment that makes access to unhealthy, dense energy foods easier and cheaper. The food industry is making society fatter everyday by providing low quality large portion sized foods that are cheaper than healthy ones (Hill, & Sallis, 2004; Cawley, 2004; Sturm, 2004). Our lifestyles are also driven by an economy that encourages less effort by providing many attractive forms of entertainment and leisure activities forcing a positive energy balance. Working behind a desk may be more productive in business achievements but at the expense of the employee's health. Presently, economic progress is fueling the obesity epidemic.

In setting the country's health-promotion and disease-prevention agenda for the next 10 years, Healthy People 2020 adds two more goals to improve the health of the nation: Promoting quality of life, healthy development, and healthy behaviors across life stages; and creating social and physical environments that promote good health. As a nation we have to move beyond the traditional health care system and seek public health leadership that engages nontraditional partners at all levels of society, to create an environment where individuals and society choose to alleviate the obesity epidemic.

Acknowledgment

The authors would like to thank Ms. Lindsay Frrero for her contribution to this article.

References

Ackard, D. M., Neumark-Sztainer, D., Story, M., & Perry, C. (2003). Overweight among adolescents: prevalence and associations with weight-related characteristics and psychological health. *Pediatrics*, 111, 67 – 74.

Annesi, J. J. & Unruh, J. L. (2008). Relations of exercise, self appraisal, mood change, and weight loss in obese women: testing propositions based on Baker and Brownell's (2000) Model. *American Journal of Medical Science*, 335 (3): 198-208.

Antony, M. M., Johnson, W. G., Carr-Nangle, R. E. & Abel, J. L. (2004). Psychopathology correlates of binge eating and binge eating disorder. *Comprehensive Psychiatry*, 35, 386-392.

Baker, C. K. & Brownell, K. D. (2000). Physical Activity and Maintenance of Weight Loss: Physiological and Psychological Mechanisms. *Human Kinetics*, 311-38.

Barofsky, I., Fontaine, K. R., Cheskin, L. J. (1998). Pain in the obese: impact on health-related quality of life. *Annals of Behavioral Medicine*, 19, 408-410.

Booth, S. L., Sallis, J. F., Ritenbaugh, C., Hill, J. O., Birch, L. L., Frank, L. D., Glanz, K., Himmelgreen, D. A., Mudd, M., Popkin, B. M., Rickard, K. A., St. Jeor, S., Hays, N. P. (2001). Environmental and societal factors affect food choice and physical activity: rationale, influences and leverage points. *Nutr Rev*, 59: S21–S39.

Borradaile, K. E., Sherman, S., Vander Veur, S. S., et al. (2009). Snacking in children: The role of urban corner stores. *Pediatrics*, 124 (5):1292-1297.

Brownell, K. D. (2004). Fast Food and obesity in Children. Pediatrics, 113:132.

Brownell, K. D., Horgen, K. B (2004). *Food Fight: the Inside Story of the Food Industry, America's Obesity Crisis, and What We Can Do about It*. Contemporary Books: Chicago, IL Butler, G. S., Vallis, T. M., & Perey, B. (1999). The obesity adjustment survey: development of a scale to assess psychological adjustment to morbid obesity. *International Journal of Obesity Related Metabolic Disorders*, 23, 505-511.

Byrne, S. M., Cooper, Z., & Fairburn, C. G. (2004). Psychological predictors of weight regain in obesity. *Behavior Research and Therapy*, 42:1341-1356.

Centers for Disease Control and Prevention (2004). Trends in intake of energy and macronutrients—United States, 1971–2000. MMWR, 53:80–2. Available from: *www.cdc.gov/mmwr/preview/mmwrhtml/mm5304a3.htm*.

Childhood Obesity Statistics (2010): *http://www.childhoodobesitystatistics.net/epidemic*.

Childhood Overweight and Obesity (2010): *http://www.cdc.gov/obesity*.

Cizza, G., Skarulis, M., & Mignot, E. (2005). A link between short sleep and obesity: building the evidence for causation. *Sleep*, 28:1217–1220.

Colles, S. L., Dixon, J. B., & O'Brien, P. E. (2008). Loss of control is central to psychological disturbance associated with binge eating disorder. *Obesity*, 16, 608-614.

Comuzzie, A. G., Allison, D. B.(1998). The search for human obesity genes. *Science*, 280:1374-1377.

Cawley, J. (2004). An economic framework for understanding physical activity and eating behaviors. *Am J Prev Med*, 27: 117–126.

Department of Health and Human Services: Centers for Disease Control and Prevention (2009). *Overweight and Obesity*.

Duffy, V. B. (2007) Variation in oral sensation: implications for diet and health. *Curr Opin Gastrenterol*, 23: 171-177 pp.

Duffy, V. B., Davidson, A. C., Kidd, J. R., Kidd, K. K., Speed, W. C., Pakstis, A. J., et al. (2004). Bitter receptor gene (TAS2R38), 6-n-propylthiourcil (PROP) bitterness and alcohol intake. *Alcohol Clin Exp Res*, 28: 1629-1637.

Duffy, V. B., Lucchina, L. A., & Bartoshuk, L. M. (2004). Genetic variation in taste: potential biomarker for cardiovascular disease risk? In: Prescott J, Tepper B, editors. Sensitivity to PROP (6-n-propylthiouracil): measurement, significances and implications. Erlangen, Germany: Merrkel Decker, Inc, pp. 197–229.

Drewnowski, A (1998). Energy density, palatability, and satiety: implications for weight control. *Nutr Rev*, 56: 347–353.

Eldredge, K. L. & Agras, W. S. (1998). Weight and shape overconcern and emotional eating in binge eating disorder. *International Journal of Eating Disorders*, 19, 73-82.

Fairburn, C. G., Doll, H. A., Welch, S. L., Hay, P. J., Davies, B. A. & O'Connor, M. E. (1998). Risk factors for binge eating disorder. *Achieves of General Psychiatry*, 55, 425-432.

Finkelstein, E. A. & Zuckerman, L. (2008). The fattening of America: how the economy makes us fat, if it matters, and what to do about it. Hoboken, NJ: John Wiley & Sons.

Fischer, R., Griffin, F., Rockey, M. A. (1966). Gustatory chemoreception in man: multidisciplinary aspects and perspectives. *Perspect. Biol. Med,* 9:549–77www. healthypeople.gov

Flegal K. M., Carroll M. D., Ogden, C. L., & Curtin, L. R. (2010). Prevalence and Trends in Obesity Among US Adults, 1999-2008. *JAMA,* 303(3):235-241.

Franco, M., Diez Roux, A. V., Nettleton, J. A., et al. (2009). Availability of health foods and dietary patterns: the multi-ethnic study of atherosclerosis. *Am J Clin Nutr*, 8:1-5.

Frayling, T. M., Timpson, N. J., Weedon, M.N., et al. (2007). A common variant in the FTO gene is associated with Body Mass Index and predisposes to childhood and adult obesity. *Science*, 316: 889-94.

Glanz, K., Basil, M., Maibach, E., Goldberg, J., & Snyder, D. (1998). Why Americans eat what they do: taste, nutrition, cost, convenience, and weight control concerns as influences on food consumption. *J. Am. Diet. Assoc.* 98:1118–26.

Gordon-Larsen P., Nelson, M. C., Page, P. & Popkin (2006). Inequality in the built environment underlies key health disparities in physical activity and obesity. *Pediatrics*, 117: 417-24.

Handy S., Cao, X. & Mokhtarian, P. L. (2008). The causal influence of neighborhood design on physical activity within the neighborhood: evidence from North Carolina. *American Journal of Health Promotion,* 22(5):350-58.

Healy, G. N., Wijndaele, K., Dunstan, D. W., Shaw, J. E., Salmon, J., and Owen, N. (2008). Objectively measured sedentary time, physical activity and metabolic risk: the Australian diabetes, obesity and lifestyle study (Australian Diabetes). *Diabetes Care*, 31, 369-371.

Hetherington, M. M. (2007). Symposium on 'molecular mechanisms and psychology of food intake' Cues to overeat: psychological factors influencing overconsumption. *Proceedings of the Nutrition Society*, 66, 113-123.

Hill, J. O. (2002). The nature of the regulation of energy balance. In: Fairburn, CG, Brownell, KD . *Eating Disorders and Obesity. A Comprehensive Handbook*, 2nd edn. The Guilford Press: New York, pp. 67–71.

Hill, J. O. & Peters, J. C. (1998). Environmental contributions to obesity epidemic. *Science*, 280:1371-1374.

Hill, J. O., Sallis, J. F., Peters, J. C. (2004). Economic analysis of eating and physical activity: a next step for research and policy change. *Am J Prev Med*, 27: 111–116.

Jakicic J. M., Otto, A. D., Lang, W., Semler, L., Winters, C., Polzien, K. & Mohr, K. I. (2010).The effects of physical activity on 18-month weight change in overweight adults. *http://www.obesityjournal.com*, Accessed 14 June 2010.

Jeffery, R. W., Wing, R. R., & Mayer, R. R. (1998). Are small weight losses or more achievable weight loss goals better in the long-term for obese patients? *Journal of Consulting and Clinical Psychology*, 66:641-645.

Klein, S., Sheard, N. F., Pi-Sunyer, X., Daly, A., Wylie-Rosett, J., Kulkarni, K., & Clark,N. G. (2004). Weight management through lifestyle modification for the prevention of type 2 diabetes: rationale and strategies. *Diabetes Care*, 27, 2067-2073.

Kruger, J., Blanck, H. M., & Gillespie, C. (2006). Dietary and physical activity behaviors among adults successful at weight loss maintenance. *International Journal of Behavior Nutrition and Physical Activity,* 3:17.

Kwon, H. R., Min, K. W., Ahn, H. J., Seok H. G., Koo, B. K., Kim, H. C., & Han, K. A. (2010). Effects of aerobic exercise on abdominal fat, thigh muscle mass and muscle strength in type 2 diabetic subject. *Korean Diabetes J.*, 34(1):23-31.

Lakdawalla, D. & Philipson, T. (2007). Labor supply and weight. *J Hum Resour*, 42:85–116.

Larsson, U., Karlsson, J., & Sullivan, M. (2002). Impact of overweight and obesity on health related quality of life- a Swedish population study. *International Journal of Obesity Related Metabolic Disorders*, 26, 417-424.

Lucan, S. C., Karpyn, A., & Sherman, S. (2010). Storing empty calories and chronic disease: snack-food products, nutritive content, and manufacturers in Philadelphia corner stores. *Journal of Urban Health: Bulletin of the New York Academy of Medicine*, 87; 3: 394-409.

MacEwen, B. S. (2000) Allostasis and allostatic load: implications for neuro-psychopharmacology. *Neuropsychopharmacology*, 22:108–124.

MacEwen, B. S. (2006). Sleep deprivation as a neurobiologic and physiologic stressor: Allostasis and allostatic load. *Metabolism*, 55(10 Suppl 2):S20–S23.

Marcellini, F., Giuli, C., Papa, R., Tirabassi, G., Faloia, E., Boscaro, M., Polito, A., Ciarapica, D., Zaccaria, M., & Mocchegiani, E. (2009). Obesity and BMI in Relation to lifestyle and psycho-social aspects. *Gerontology*, 195-206.

Marti, A., Corbalan, M. S., Martinez-Gonzalez, M. A., Forga, L. & Martinez, J. A. (2002). CHO intake alters obesity risk associated with Pro12Ala mutation of the PPARgamma gene. *J Physiol Biochem*, 58:219-220.

Marti, A., Corbalan, M. S., Martinez-Gonzalez, M. A. & Martinez, J. A. (2003). Trp64Arg polymorphism in the beta-3-adrenoceptor and obesity risk: effect modification by a sedentary lifestyle. *Diabetes Obes Metab*, 4: 28-30.

Martinez, J. A., Corbalan, M. S., Sanchez-Villegas, A., Forga, L., Marti, A., & Martinez-Gonzalez, M. A. (2003). Obesity risk associated with carbohydrate intake in women carrying the Gln27Glu polymorphism. *J Nutr*, 2549-2552.

Mayer-Davis, E. J. & Costacou, T. (2001). Obesity and sedentary lifestyle: modifiable risk factors for prevention of type 2 diabetes. *Current Diabetes Reports*, 1, 170-176.

Mokdad, A. H., Bowman, B. A., Ford, E. S., Vinicor, F., Marks, J. S., & Koplan, J. P. (2001). The continuing epidemics of obesity and diabetes in the United States. *Journal of the American Medical Association*, 286, 1195-200.

Morland, K., Diez Roux, A. V., Wing, S. (2006). Supermarket, other food stores and obesity; the atherosclerosis risk in communities study. *Am J Prev Med*, 30: 333-9.

Morton, L. W., & Blanchard, T. C. (2007). Starved for Access: Life in Rural America's Food Deserts. *Rural Realities*, 1:1-10.

Mussell, M. P., Mitchell, J. E., de Zwaan, M., Crosby, R. D., Seim, H. C. & Crow, S. J. (1996) Clinical characteristics associated with binge eating in obese females: a descriptive study. *International Journal of Obesity*, 20, 377-388.

Mustillo, S., Worthman, C., Erkanli, A., Keeler, G., Angold, A., & Costello, E. J. (2003). Obesity and pediatric disorder: developmental trajectories. *Pediatrics*, 111, 851-859.

Ogden, C. L., Carroll, M. D., & Flegal, K. M. (2008). High body mass index for age among US children and adolescents 2003- 2006. *Journal of the American Medical Association*, 299, 2401-2405.

Ogden, C. L., Yanovski, S. Z., Carroll, M. D., & Flegal, K. M. (2007). The Epidemiology of Obesity. *Gastroenterology*, 132:2087-2102.

Orbach, S. (2002). *On eating*. London, UK: Penguin.

Ponder, S. W., & Anderson, M. A. (2008) Teaching families to keep their children s.a.f.e. from obesity. *Diabetes Spectrum*, 21, 50-53.

Popkin, B. M. (2008). *The world is fat—the fads, trends, policies, and products that are fattening the human race*. New York, NY: Avery-Penguin Group.

Powell, L., Auld, C., Chaloupka, F., O'Malley, P., & Johnston L. (2007). Associations between access to food stores and adolescent body mass index. *Am J Prev Med*, 33:S301-S307.

Risks, P. (2010) Tips for parents – ideas to help children maintain a healthy weight. *CDC, cdc.gov/healthyweight/children*.

Roberts, D., Foehr, U. & Rideout, V. (2005). *Generation M: media in the lives of 8 to 18 year-olds*. Menlo Park, CA: The Henry J Kaiser Family Foundation.

Rodriguez-Baziuk, V. (2009). Increase serotonin and endorphins naturally. *NIMH*: Healthpsychology.suite101.com .

Schwimmer, J.B., Burwinkle, T.M., and Varni, J. W. (2003). Health related quality of life of severely obese children. *Journal of the American Medical Association*, 289, 1813-1819.

Specker, S., de Zwaan, M., Raymond, N. & Mitchell, J. (2004). Psychopathology in subgroups of obese women with and without binge eating disorder. *Comprehensive Psychiatry*, 35, 185-190.

Sonestedt, E., Roos, C., Gullberg, B., Ericson, U., Wirfalt, E. & Orho-Melander, M. (2009). Fat and Carbohydrate intake modify the association between genetic variation in the FTO genotype and obesity. *Am J Clin Nutr*, 90: 1418-25.

Song, S. H., & Hardisty, C. A. (2008) Early-onset type 2 diabetes mellitus: an increasing phenomenon of elevated cardiovascular risk. *Expert Review of Cardiovascular Therapy*, 6, 315-322.

Steiner, J. E., Glaser, D., Hawilo, M. E., & Berridge, K. C. (2001). Comparative expression of hedonic impact: affective reactions to taste by human infants and other primates. *Neurosci Biobehav Rev*, 25: 53-74.

Sturm, R. (2004). The economics of physical activity: societal trends and rationales for intervention. *Am J Prev Med*, 23(Suppl. 3): 136–145.

Tanofsky-Kraff, M., Han, J. C., Anandalingam, K., Shomaker, L. B., Columbo, K. M., Wolkoff, L. E., Kozloskym, M., Elliot, C., Ranzenhofer, L. M., Roza, C. A., Yanovski, S. Z., & Yanovski J. A. (2009) The FTO gene rs9939609 obesity risk allele and loss of control overeating. *Am J Clin Nutr*, 90: 1483-8.

Tarasuk, V., & Beaton, G. H. (1991). The nature and individuality of within-subject variation in energy intake. *Am J Clin Nutr*, 54: 464–470.

The American Physiology Society. Obesity. *American Obesity Association website: http://www.obesity*

The Writing Group for the SEARCH for Diabetes in Youth Study Group. (2007) Incidence of diabetes in youth in the United States. *Journal of the American Medical Association*, 297, 2716-2724.

Tiexiera, P. J., Going, S. B., Hout Kooper, L. B., Cussler, E. C., Martin, C. J., Metcalfe, L., et al. (2002). Weight Loss readiness in middle-aged women: psychological predictors of success for behavioral weight reduction. *J Med*, 25:499-523.US Department of Health and Human Services (2008). Physical Activity Guidelines for Americans. Washington DC.

Wadden, T. A. & Phelan, S. (2002). Assessment of quality of life in obese individuals. *Obesity Research*, 10, 50S-57S.

Wilfley, D. E., Schwarts, M. B., Spurrell, E. B. & Fairburn, C. G. (2000). Using the eating disorder examination to identify the specific psychopathology of binge eating disorder. *International Journal of Eating Disorders*, 27, 259-269.

Young, L. R. & Nestle, M. (2002). The contribution of expanding portion size to the US obesity epidemic. *American Journal of Public Health*, 92, 246-249.

Young, S. N. (2008). The neurobiology of human social behaviour: an important but neglected topic. *Journal of Psychiatry and Neuroscience*, 33, 391-392.

In: Childhood Obesity: Risk Factors, Health Effects...
Editor: Carol M. Segel

ISBN: 978-1-61761-982-3
©2011 Nova Science Publishers, Inc.

Chapter III

The Prevalence of Overweight and Obesity and Related Metabolic Abnormalities among Children and Adolescents in China

Zhijie Yu[*1], Gang Hu[2], and Xilin Yang[3]*
[1]Institute for Nutritional Sciences
Shanghai Institutes for Biological Sciences,
Chinese Academy of Sciences, Shanghai, China;
[2]Pennington Biomedical Research Center, Baton Rouge, LA, USA
[3]The Chinese University of Hong Kong, Hong Kong, SAR, China

Abstract

Along with the rapid economic and social development during the past three decades, China is experiencing a dramatic epidemiologic transition. Noncommunicable chronic diseases including cardiovascular disease, cancer, and type 2 diabetes mellitus have been becoming the leading cause of death and disability among Chinese people. Obesity plays an important role in the pathogenesis of cardiometabolic disease. There is a growing body of evidence suggesting that overweight or obesity in childhood are associated with an increased risk of being overweight or obese and having obesity related metabolic disorders during adulthood. Few studies have systematically assessed changes in the prevalence of overweight and obesity, the health consequences, and related risk factors among Chinese children and adolescents. In this review, we attempt to summarize: 1) the prevalence of overweight and obesity and its time trends among Chinese children younger than 18 years of age; 2) overweight and obesity related metabolic abnormalities; and 3) associated risk factors.

* Corresponding author: Zhijie Yu, MD, PhD, MPH, Key Laboratory of Nutrition and Metabolism, Institute for Nutritional Sciences, Shanghai Institutes for Biological Sciences, Chinese Academy of Sciences, Shanghai, China, 200031, Tel: +86 21 54920902, Fax: +86 21 54920291, Email: zjyu@sibs.ac.cn

Key words: overweight, obesity, trends, children, adolescents, metabolic disorder, China

1. Introduction

China is experiencing a remarkable epidemiologic transition along with her rapid economic and social development during the past three decades [1]. Noncommunicable chromic diseases, including cardiovascular disease (CVD), cancer, and type 2 diabetes mellitus (T2D), have been becoming the leading cause of death and disability among Chinese people, especially for populations living in urban areas [2, 3]. Overweight and obesity are major risk factors of cardiometabolic disease, such as CVD, T2D, and certain cancers [4]. The prevalence of overweight and obesity and related metabolic disorders have been reaching an alarming rate among Chinese living in mainland China [5-8]. Data from the China National Nutrition and Health Survey 2002 showed that there are 281 million people who are overweight or obese according to the Chinese criteria (body mass index [BMI; kg/m^2] $\geq$ 24). Defining overweight and obesity with the World Health Organization (WHO) criteria, i.e., BMI $\geq$ 25, the number is 215 million [7]. A large-scale population-based study reported that the prevalence of metabolic syndrome is 16.5% among Chinese adults 35-74 years of age by the International Diabetes Federation (IDF) criteria. If it is defined with the National Cholesterol Education Program-Adult Treatment Panel III (NCEP-ATPIII) criteria, the prevalence rate is 23.3% [9]. During 2007-2008, a national survey on diabetes mellitus was carried out among Chinese adults 20 years of age or older living in mainland China. The prevalence of diabetes mellitus was 9.7% and another 15.5% was prediabetes. It has been estimated that there are about 92.4 million adults with diabetes and another 148.2 million adults with prediabetes [8]. The situation is even worse among big cities and elderly populations. Findings from the Nutrition and Health of Ageing Populations in China Study indicate that the prevalence of diabetes mellitus is 13.6% among people 50-70 years of age living in Beijing and Shanghai, China. Meanwhile, the prevalence of the metabolic syndrome is 42.4% according to the NCEP-ATPIII criteria [10]. Currently, China is facing to a tide of obesity and related cardiometabolic disease.

There is a growing body of evidence that children who are overweight or obese during childhood are associated with increased risks of being overweight or obese as well as developing obesity related cardiometabolic disease in adulthood [11,12]. Better understanding of the prevalence of overweight and obesity and associated metabolic abnormalities among children and adolescents in China may help predict the burden of obesity and related cardiometabolic disease among adults in the future. It may also trigger more research activities and policy makings to counteract the upcoming epidemic of obesity and related metabolic diseases.

In this review, we attempt to summarize research findings on the prevalence of overweight and obesity and obesity related metabolic disorders among children and adolescents in China. Moreover, we summarize risk factors associated with development of obesity and related metabolic disorders as well. We focus on studies carried out in the mainland and Hong Kong, China and published in English. Finally, we propose some implications for future studies in Chinese children.

2. Prevalence of Overweight and Obesity among Children and Adolescents in China

The prevalence of overweight and obesity in children and adolescents varies substantially among age, gender, geographic region and residence in China. The different definitions of overweight and obesity adopted in different studies may contribute to the variation and make the difficulties for direct comparison among different studies (Tables 1 and 2). Several definitions were adopted in the summarized studies. The most frequently adopted references are the International Obesity Task Force and the Working Group on Obesity in China.

The International Obesity Task Force criteria are age and sex-specific BMI cutoff points derived from sex-specific age curves that pass through a BMI of 25 and 30 at 18 years of age [13]. The BMI cutoff points are the WHO criteria of overweight and obesity for adults. The reference was developed according to the data collected in Brazil, Britain, Hong Kong, the Netherlands, and the US.

The definitions of overweight and obesity of the Working Group on Obesity in China are age and sex specific BMI cutoff points derived from sex-specific age curves that pass through a BMI of 24 and 28 at 18 years of age [14]. The BMI cutoff points are the Chinese criteria of overweight and obesity for adults. The reference was developed based on data from the Chinese National Survey on Students Constitution and Health 2000.

There are some concerns that the WHO criteria of overweight and obesity may underestimate the prevalence of overweight and obesity among Asian populations, given that Asians have greater amount of total fat mass than Europeans with the same BMI [15]. With a national representative sample of mainland China, a study in this regard found that the prevalence rate of overweight and obesity was 17.3% as per the WHO criteria. Whereas, the prevalence rate increased by about 6% (23.2%) with the Chinese criteria. Totally, the number of individuals who were overweight or obese would be decreased by 66 million when the WHO criteria were adopted among Chinese people [7].

2.1. Distribution of Prevalence of Childhood Overweight and Obesity

No matter which criteria are adopted to define overweight or obesity among Chinese children and adolescents, it seems likely that the prevalence rate of overweight, in most of the studies, is higher than the prevalence of obesity (Table 1). There is a geographic difference in the prevalence of overweight and obesity among children and adolescents in mainland China. The prevalence rate is higher among those living in the north than those in the south of China. Data from a national survey of school-aged children and adolescents (7-18 years of age) showed that the prevalence of overweight and obesity was 32.5% among boys and 17.6% among girls in north coastal big cities. Whereas, the corresponding percentages were 20.1% and 10.3%, respectively, among their counterparts in south coastal big cities [16]. Liu and coworkers [17] evaluated the prevalence of overweight and obesity among 262,738 children 3-6 years of age in 26 counties and cities in mainland China. The study included five counties in one northern province, and 15 counties and five cities in two southern provinces. They defined overweight and obesity with the International Obesity Task Force criteria. Compared with children living in southern cities, those from the north had a likelihood of 2.58 (95%

confidence interval [CI], 2.43-2.73) to be overweight or obese with adjusted for age, gender, residence area, maternal age at delivery, occupation, education, and ethnicity.

Table 1. Prevalence of overweight and obesity among children and adolescents in China

Study (year of survey)	Age, y	Setting	Criteria	Prevalence, %		
				Overweight	Obesity	Overweight and obesity
Wang (1993) [19]	6-9	8 provinces in mainland China	WHO/NCHS reference	4.6	7.3	11.9
	10-18			3.0	1.8	4.8
Hui and Bell (1999) [24]	7-12	A representative sample in Shenzhen, China	International Obesity Task Force	12.4	3.2	15.6
Liu et al. (2000) [32]	3-6	Five counties in Hebei province and sixteen counties and five cities in Zhejiang province and Jiangsu province, China	International Obesity Task Force	6.4	1.0	7.4
Wu (2002) [7]	0-6	National Nutrition and Health Survey in mainland China	0-6: WHO criteria 7-17: overweight ≥ 85th centile; obesity ≥ 95th centile	3.4	2.0	5.4
	7-17			4.5	2.1	6.6
Shi et al. (2002) [26]	12-14	A representative sample of middle schools in two cities in Jiangsu province, China	WHO/NCHS reference	8.7	4.9	13.6
Kong et al. (2003) [33]	11-16	A representative sample of secondary schools in Hong Kong, China	Hong Kong reference	12.0	6.2	18.2
Xu et al. (2004) [25]	12-18	A representative sample of senior and junior high schools in Nanjing, China	Working Group on Obesity in China	NA	NA	13.2
Zhou et al. (2004) [23]	9-16	One primary school and one middle school in Dalian, China	US-CDC reference	NA	NA	16.6
Ji and Cheng (2005) [16]	7-18	Chinese National Survey on Students' Constitution and Health	Working Group on Obesity in China	8.0	3.8	11.8
Zhang et al. (2005) [34]	3-6	A representative sample of kindergartens in Tianjin, China	WHO child growth reference	9.8	8.0	17.8

NA = not available.

Currently, there seems to be an opposite tendency on the association between socioeconomic development and the prevalence of overweight or obesity among Chinese children as compared to data from Western developed countries in which an inverse association of socioeconomic development with incidence of childhood overweight and obesity is observed [18, 19]. In China, the prevalence of childhood overweight and obesity is higher among children and adolescents living in areas with relatively higher socioeconomic development than those in the underdeveloped areas. Data from the China Nutrition and Health Survey 2002 showed that the prevalence of overweight and obesity was 9.8% and 1.7%, respectively, among boys aged 7-17 years in urban areas [20]. For girls, the corresponding percentages were 7.1% and 0.9%, respectively. While, in rural areas, the prevalence of overweight and obesity was 3.3% and 0.9 among boys, and 2.9% and 0.5% among girls, respectively. Ji and Cheng [16] analyzed the largest, so far, nationally representative sample of school children and adolescents (7-18 years of age) in mainland China. The sample was drawn from 30 of the 31 provinces and municipalities in 2005. They reported that among boys, the prevalence of overweight and obesity was 19.3% and 13.2% respectively, in the north coastal big cities, followed by 14.1% and 6.0% in the south coastal big cities, and 13.1% and 7.1% in the inland big cities. In Girls the corresponding percentages were 10.8% and 6.8% in the north coastal big cities, 7.4% and 2.9% in the south coastal big cities, and 7.4% and 3.3% in the inland big cities, respectively. The prevalence of overweight and obesity was relatively lower in the eastern and middle rural areas. The rates were 7.3% and 3.3% among boys, and 5.4% and 1.9% among girls, respectively. In the western rural areas the prevalence rate was among the lowest. In boys, the prevalence of overweight and obesity was 3.7% and 1.3%, respectively. The corresponding percentages were 3.1% and 0.8%, respectively, among girls.

Consistently, the prevalence of overweight and obesity is higher among boys than in girls. In two national representative samples, the proportion of boys who are overweight or obese is higher than that of girls [16, 20]. This seems to be true across different age groups [20], geographic region and residence as well as areas with different socioeconomic development [16]. Data from studies carried out in various settings provide strong evidence. A multi-city comparison study indicated that the percentages of children 0-19 years of age with a BMI more than 95th centile was higher among boys than girls [21]. This trend is consistent across the four cities involved in the study. In Chinese Hong Kong, data from two representative samples of children aged 6-18 years in different survey years showed the similar pattern [22]. In 1993, the prevalence of overweight and obesity was 13.8% in boys and 9.5% in girls, respectively. In 2005, the prevalence rate increased to 13.5% among girls but was still lower than that among boys which was 20.9%. In a northern city of Dalian, the prevalence of overweight and obesity in boys was more than twice as that in girls (boys vs. girls; 22.9% vs. 10.4%) [23]. In Shenzhen, boys 7-12 years of age had a relative risk of 1.92 (95% CI, 1.62-2.27) to be overweight or obese compared with girls [24]. Similar findings were observed in Nanjing, the capital city of Jiangsu Province, where among children 12-18 years of age, boys had about a two-fold increased risk (odds ratio [OR], 2.12; 95% CI, 1.74-2.60) of being overweight or obese than girls [25]. In another survey carried out in the province of Jiangsu, the proportion of overweight and obesity was also higher among boys (17.9%) than girls (8.9%) 12-14 years of age [26].

2.2. Increasing Trends in the Prevalence of Childhood Overweight and Obesity in China

During the past three decades, the prevalence of overweight and obesity among children and adolescents increased greatly in China (Table 2). Data from the nutrition surveillance program and the National Child Survey reported that infants (0-6 years of age) with a weight-for-height z score > 2 raised from 3.5% in 1990 to 6.7% in 1995 in urban areas. Among infants in rural areas the increase was even greater, i.e., from 1.1% in 1990 to 12.6% in 1995 [27]. Similar trends were observed among school-aged children and adolescents (7-18 years of age) during 1985-2000 in the Chinese National Survey on Students' Constitution and Health [28]. In the coastal big cities, the percentages of overweight and obesity increased from 1.8% in 1985 to 18.6% in 2000 among boys, and from 1.8% to 13.6% among girls, respectively. In the inland middle and small cities, the prevalence rate was increased from 1.2% in 1985 to 7.2% in 2000 among boys, and from 2.0% to 6.3% among girls, respectively. A study of Li et al [20] analyzed the trends in the prevalence of overweight and obesity among children 7-17 years of age with the data obtained from three rounds of the China Nutrition and Health Survey in each 10-year interval from 1982 through 2002. By using the

International Obesity Task Force criteria, they reported that the prevalence of overweight and obesity was increased four times during the 20 years (from 1.3% to 5.2%). A multi-national comparison study [29] indicated that mean BMI of obese Chinese children 6 years of age is 2.6 units higher than their US counterparts in 2006.

Table 2. Trends in prevalence of overweight and obesity among children and adolescents in China

Study (year of publication)	Age, y	Setting	Criteria	Time period	Prevalence, %		
					Overweight	Obesity	Overweight and obesity
Chen (2000) [27]	0-6	Nutrition surveillance and National Child Survey	Weight-for-height z score > 2	Urban areas			
				1990	NA	3.5	NA
				1992	NA	3.8	NA
				1995	NA	6.7	NA
				Rural areas			
				1990	NA	1.1	NA
				1992	NA	10.4	NA
				1995	NA	12.6	NA
Luo and Hu (2002) [30]	2-6	7 provinces in mainland China	International Obesity Task Force	1989	NA	4.2	NA
				1997	NA	6.4	NA
Wang et al. (2002) [31]	6-18	8 provinces in mainland China	International Obesity Task Force	1991	NA	NA	6.4
				1997	NA	NA	7.7
Ji (2008) [28]	7-18	Chinese National Survey on Students' Constitution and Health	Working Group on Obesity in China	Boys in costal big cities			
				1985	1.5	0.3	1.8
				1991	5.8	1.8	7.6
				1995	9.9	4.5	14.4
				2000	14.5	4.1	18.6
				Girls in costal big cities			
				1985	1.6	0.2	1.8
				1991	2.8	1.4	4.2
				1995	5.9	2.8	8.7
				2000	8.8	4.8	13.6
So et al. (2008) [22]	6-18	Representative surveys in Hong Kong, China	International Obesity Task Force	1993	9.0	2.6	11.6
				2005/6	13.0	3.7	16.7
Li et al. (2008) [20]*	7-17	China National Nutrition Survey	International Obesity Task Force	1982	1.2	0.2	1.3
				1992	3.7	0.9	4.4
				2002	4.4	0.9	5.2

NA = not available. *Standardized rates according to the population structure of the Fifth National Population Census

The increasing trends are also observed in the studies carried out in some local areas in China. Lu and Hu [30] reported that infant obesity (2-6 years of age) was 4.2% in seven provinces in 1989, but increased to 6.4% in 1997. Similarly, Wang and colleagues [31] found that percentages of overweight and obesity among children 6-18 years of age increased from 6.4% in 1991 to 7.7% in 1997 in the same survey. In Hong Kong, the prevalence of overweight and obesity raised from 11.6% in 1993 to 16.7% in 2005/6 [22].

In summary, the prevalence of overweight and obesity among children and adolescents vary substantially among geographic region and residence, socioeconomic development, and gender. There is convincing evidence that childhood overweight and obesity have been increasing rapidly across the whole country. In most of the well developed cities in mainland China, the prevalence rate is almost identical to that of Hong Kong. It seems very likely that the epidemic of childhood overweight and obesity among Chinese children and adolescents and its health consequence have been becoming a critical public health issue in China.

3. Obesity Related Metabolic Abnormalities

Chinese may be "metabolically obese", given that Chinese has greater amount of abdominal fat mass than Caucasians with the same total amount of body fat mass [35]. However, how does the increasing of the prevalence of overweight and obesity among Chinese adults contribute to the morbidity and mortality from cardiometabolic disease has not yet been fully assessed [36]. For children and adolescents, only during recent years, have limited studies reported childhood obesity related metabolic disorders in China. Therefore, studies summarized here may only provide some implications for studies in the future.

In the China Nutrition and Health Survey, anthropometric indexes, blood pressure, and blood sample were measured among 8861 children and adolescents 7-18 years of age. Overweight and obesity were defined according to the Chinese criteria. All metabolic traits in terms of waist circumference, blood pressure, total cholesterol, low-density lipoprotein (LDL) cholesterol, high-density lipoprotein (HDL) cholesterol, triglycerides, and fasting glucose, were significantly higher among those who were overweight or obese than those with a normal weight [37]. Overall 3.3% of children and adolescents had the metabolic syndrome according to the definition for US adolescents. 58.3% of the study participants who had the metabolic syndrome were either overweight or obese. Participants who were overweight or obese were 15 times more likely of having the metabolic syndrome compared with their normal weight counterparts.

The prevalence of diabetes mellitus was 0.2% among children 7-11 years of age and was 0.4% among those 12-18 years of age. Compared with children with a normal weight, overweight and obese children were two times more likely (OR, 2.3; 95% CI, 1.00-5.4) of having impaired fasting glucose. The relative risks were more than 1.5 and three times for childhood dyslipidemia and hypertension, respectively.

Similar findings are reported among children and adolescents in Hong Kong, China. Sung and coworkers [38] compared metabolic traits between 129 obese and 142 nonobese children 9-12 years of age. They found that about 50% obese children had more than two of the three cardiometabolic risk factors including dyslipidemia, elevated blood pressure, and hyperinsulinemia. 8% had all these three risk factors. Importantly, more than 77% of obese children had hyperinsulinemia indicating increased risk for future T2D development. Compared with nonobese children, those who were obese had increased ORs of 3.21 (95% CI, 1.60-6.45), 2.72 (95% CI, 1.58-4.66), and 14.1 (95% CI, 17.75-25.48) of having high blood pressure, dyslipidemia, and hyperinsulinemia, respectively. In a random sample of 2102 adolescents 11-16 years of age from 14 secondary schools in Hong Kong, Kong and colleagues [33] reported that those who were overweight or obese had significantly higher levels of not only metabolic traits, i.e., waist circumference, waist-to-hip ratio, blood pressure, HDL-cholesterol, LDL-cholesterol, triglycerides, fasting glucose, and HOMA-IR, but also higher levels of sub-clinical chronic inflammation markers, i.e., high-sensitivity C-reactive protein (hsCRP) and white blood cell count. A growing body of evidence has suggested that sub-clinical chronic inflammation is one of the major culprits linking obesity and cardiometabolic disease [39, 40].

4. Risk Factors Associated with Overweight or Obesity and Related Metabolic Abnormalities among Chinese Children and Adolescents

Risk factors associated with development of childhood obesity and related metabolic disorders are of increasing concerns during recent years. The fetal and infant origin of adult disease hypothesis [41] raises broad spectra of imaginations on the searching for risk factors related to fetal and infant programming of adult disease. The risk factors associated with childhood obesity and related metabolic abnormalities may have profound influence on adult cardiometabolic disease development. There is a dramatic increase of research findings in the international society from various disciplines. Of them, findings based on studies carried out in China may provide some explanations on the increasing trends in childhood overweight and obesity among Chinese children and adolescents as well as some implications for future studies.

4.1. Prenatal Factors

Gestational diabetes mellitus (GDM) is an important risk factor of newborn macrosomia, childhood obesity, and obesity related metabolic abnormalities in children and adolescents [42-44]. Yang and coworkers [45] reported that Chinese pregnant women with GDM were two times more likely (OR, 2.42; 95% CI, 1.07-5.46) to delivery a baby with a high birth weight (more than the 90th centile or 4000 g) compared with pregnant women with euglycemia.

Two studies of Tam and colleagues [46, 47] examined the relationships of exposure to maternal gestational diabetes mellitus with cardiometabolic risk factors among children and adolescents in Hong Kong, China. Among children 7-10 years of age, those who exposed to maternal diabetes mellitus had significantly higher blood pressure levels and lower HDL cholesterol concentrations than those without exposure to maternal diabetes [46]. In another study [47], they assessed levels of cardiometabolic traits between adolescents 15 years of age with and without exposure to maternal GDM. Adolescents who exposed to maternal hyperinsulinemia had increased risks of 10.8 (95% CI, 1.69-69.2) and 17.6 (95% CI, 1.32-235), respectively, to be overweight or obese and to have the metabolic syndrome.

4.2. Birth Weight and Infant Feeding Modalities

Both birth weight and infant feeding modalities are important factors associated with childhood obesity development [48-52]. It apears to be true in both Caucasians [52] and Asians [50]. However, the postpartum environment, especially infant feeding modalities, may play a more important role in childhood obesity development [53, 54].

In a retrospective cohort study [55], we analyzed the relationships between high birth weight and childhood overweight and obesity among 918 infants 1-3 years of age living in an urban district of Shanghai, China. High birth weight was defined as birth weight $\geq 90^{th}$ percentile of sex-specific birth weight distribution. We calculated the weight-for-length/height z score for each infant with parameters of the WHO child growth reference. Childhood overweight and obesity were defined as a weight-for-length/height z score ≥ 1.68. High birth weight was associated with an odds ratio of 2.33 (95% CI, 1.29-4.22) to develop childhood overweight or obesity with controlling for age, gender, illness status, the age of breast feeding cessation, and the age at the introduction of complementary foods. In the sex stratified multivariable regression analyses, the significant association between high birth weight and childhood overweight and obesity was only apparent among boys but not girls. Boys who were macrosomia at birth and were introduced to cereals at the age of less than six moths were eight times more likely to be overweight or obese compared with those who were nonmacrosomia and were introduced to cereals at the age of more than six months.

In another retrospective longitudinal study of about 16000 children 3-6 years of age in Tianjin, China [34], we analyzed the associations between birth weight and overweight and obesity with adjusted for age, gender, gestational age, past and current health status, parents' education, occupation and health status, and family income. Overweight and obesity were defined as a BMI-for-age z score ≥ 1.65 according to the WHO child growth reference. Compared with children having a birth weight of 2500-2999 g, those with a birth weight of 3000-3499 g, 3500-3999 g, and ≥ 4000 g had significantly increased risks of developing overweight or obesity. The corresponding ORs (95% CIs) were 1.58 (1.33-1.88), 2.09 (1.76-2.49), and 3.14 (2.60-3.79), respectively. Given that about 88% of our study participants had a birth weight within the range of 3000 to 3999 g, which has been previously considered as having a lower risk of cardiometabolic disease development during adulthood. We hypothesized that the universally increased risks of developing overweight or obesity among children with the normal birth weight might be attributable to the interaction between intrauterine development and the obesogenic environment. The role of environment might be substantial. It should be pointed out that diet and nutrient intake and physical activity were not taken into account in these analyses leaving a gap for relatively fully assessing the contribution of environmental factors to childhood overweight and obesity development among children in China.

4.3. Diet and Nutrition

A rapid nutritional transition has been experiencing among populations in mainland China [56, 57]. During the past three decades, the major changes in food and nutrient intakes among Chinese are the significantly increased consumption of animal foods and plant oil, and the decreased intake of carbohydrates [58, 56, 59, 57]. Meanwhile, more foods with high energy density are taken by individuals who are socioeconomically privileged, especially for those living in the urban areas [60, 61]. These seem to be also the case for children. Data from the China Health and Nutrition Survey (CHNS) indicated that proportion of energy from fat was about 17% in 1989 among boys 12-15 years of age. It was increased to 30% in 1993- a 76% increase during four years. Children from families with higher socioeconomic status are more likely to consume snacks and processed foods than those from the socially deprived families [62]. Li and colleagues [63] analyzed major nutrients intakes between normal weight and overweight or obese children 7-17 years of age who participated in the China Nutrition and Health Survey 2002. They found that children and adolescents who were overweight or obese had significantly higher intakes of total energy, protein, and fat but less intake of carbohydrate.

4.4. Lifestyles

Physical activity and television viewing are two important factors associated with overweight and obesity among Chinese children and adolescents. Data from the CHNS seem to be optimistic with regards to physical activities among Chinese children and adolescents. About 84% of school-aged children commute to school by foot or by bicycle for about 100-150 min/week. Approximately 72% of study participants engage in moderate to vigorous activities in school for about 90-110 min/week. 72% of subjects engage in physical activities outside of school for a median of 420 min/week. Only 8% of study participants watch television for more than two hours per day [64]. However, findings from some studies are of concern with respect to associations between physical activity and obesity risk among Chinese children and adolescents.

A study of Li et al [63] reported that overweight or obese children engaged in less moderate to vigorous activities and had a longer duration of inactivity time than their normal weight counterparts. Ma and coworkers [65] analyzed the association of television viewing with the prevalence of obesity among 9356 children and adolescents. They found that each a 1-hour increment of television viewing was related to 1-2% increase in the prevalence of obesity in China. In the city of Nanjing, Xu et al [25] investigated the association between television viewing with risk of overweight or obesity among 6848 children 12-18 years of age. Compared with children who watched television for less than seven hours per week, those who watched television for more than seven hours per week had a likelihood of 1.38 (95% CI, 1.08-1.78) of having overweight or obesity with controlling for age, gender, residence, other activity time, and monthly pocket money.

Besides physical activity and television viewing, other behavioral factors may also involve in the pathogenesis of childhood obesity and obesity related metabolic disorders. There is a growing body of evidence suggesting that short sleep duration is associated with increased likelihood of having overweight or obesity among children and adolescents [66, 67]. It seems to be also likely that short sleep duration plays an important role in the pathogenesis of obesity related metabolic abnormalities [68, 69]. A Study carried out among 500 Chinese adolescent twins showed that short sleep duration (eight hours or less/day) was associated with higher levels of BMI and total body fat mass than those with relatively adequate sleep duration (nine hours or more/day) [70]. The associations seemed to be stronger for anthropometric indexes of central obesity in terms of waist circumference and truncal fat mass.

In an age-matched case-control setting of 619 obese and 617 nonobese Chinese infants 3-6 years of age, we investigated the association between parents' reported sleep duration and hyperglycemia [71]. We found that the proportion of obese children who had short sleep duration (eight hours or less/per night) were about 10% higher than their nonobese counterparts (47% vs. 36.8%, $P < 0.001$). Compared with those who slept for nine or ten hours per night, those who slept for eight hours or less had a significantly higher likelihood of having hyperglycemia (OR, 1.64; 95% CI, 1.09-2.46) after controlling for age, gender, children's anthropometric indexes, blood pressure, infant feeding modalities, current disease status, habitual food intake, physical activity, and parents' socioeconomic status and BMI. In the stratified multivariable regression analyses, infants who were overweight or obese and slept for eight hours or less had a two-fold increased risk (OR, 2.12; 95% CI, 1.06-4.21) of having hyperglycemia compared with those who were nonobese and slept for nine hours or more.

These findings suggest the need of further studies to investigate factors influencing the formation of childhood behaviors in relation to development of obesity and related metabolic disorders.

5. Nutrition and Lifestyle Modifications on Childhood Obesity Control

Thus far, nutrition and lifestyle intervention has been proven to be the most effect means for the control and prevention of obesity and related metabolic disease [72, 73]. However, well-designed large-scale intervention studies among children and adolescents are scanty.

In a school of Beijing, China, 68 obese children participated in a two-year randomized behavioral intervention study [74]. Of them, 33 children were assigned into the treatment group and 35 were assigned into the control group, respectively. Behavioral interventions were targeted to reduce total energy intake and adopt a balanced diet as well as promoting to be physically activity. There was no corresponding recommendation to participants in the control group. After the two-year intervention, mean BMI reduced by 2.6 (95% CI, -2.06 to -3.18) in the treatment group and remained almost unchanged in the control group. Meanwhile, levels of systolic and diastolic blood pressure, total cholesterol, and triglycerides were significantly decreased among intervention group compared with control group (all $P < 0.01$).

In the city of Hefei, China, a kindergarten-based intervention study was carried to test whether nutritional and behavioral intervention might improve infants and their parents' attitude, knowledge, and food consumption toward a healthy diet and food selection habit with respect to overweight and obesity development [75]. Four kindergartens with 1252 children were randomly assigned to the intervention group and three kindergartens with 850 children were assigned to the control group. Both children and their parents in the intervention group accepted monthly nutritional education during a one-year period. The dietary behavior education led to significantly improved attitude, knowledge and dietary habit in the intervention group compared with the control group. However, differences in anthropometric indexes, i.e., weight, height, and z scores of weight-for-age and height-for-age were not statistically significant between the two groups.

Currently, the China CDC is implementing a two-year large-scale randomized intervention trial involving 9750 students 7-13 years of age in six municipalities and provinces in China from 2009 through 2010 [76]. This is a multi-center study with six primary schools being selected and randomly assigned into the intervention and control groups in each center. Major intervention programs are nutritional education to improve healthy food intake and "Happy 10" program [77] to increase levels of physical activities in school or at home. Main outcomes are changes in anthropometric indexes, body composition, markers on glucose and lipid metabolism, foods and nutrients intake, levels of physical activity, and health knowledge. Meanwhile, cost-effectiveness analysis will be conducted as well.

6. Conclusions and Implications for Furture Studies

In summary, the prevalence of overweight and obesity among children and adolescents is increasing dramatically in China during the past three decades. In the areas with advanced economic and social development, i.e., eastern coastal cities, and the big cities in the north of China, the prevalence rate has reached to an alarming level. Boys are more likely than girls to be overweight or obese. The situation seems to be even worse among infants 6 years or younger. These data might indicate an upcoming epidemic of overweight and obesity among the young generation in China in the near future. This inference is made based on the rapid economic development and changes in lifestyles among contemporary Chinese populations. More importantly, these data underscore the need of appropriate intervention strategies to curtail the increasing trends in obesity and related metabolic disorders in China.

It is also true that overweight or obese children and adolescents are associated with an increased risk of having various metabolic disorders, including altered glucose and lipid metabolism as well as sub-clinical chronic inflammation. The obesity and related metabolic abnormalities during childhood might suggest the potentially increased cardiometabolic disease risk in adulthood.

Affluent diets and sedentary lifestyles appear to be the major contributions to the increasing of obesity and metabolic disorders among children and adolescents. Additionally, the substantial role of intrauterine development in the pathogenesis of childhood obesity and related metabolic abnormalities is of special concern, given that more than 200 million Chinese adults are overweight or obese as well as having impaired glucose metabolism. Postnatal factors such as infant feeding modalities and sleeping behavior seem to be involved in the pathophysiological procedure as well. More studies are warranted to elucidate prenatal and postnatal factors contributing to the pathogenesis of childhood obesity and related metabolic disorders.

There are some reasons to have an optimistic expectation as some intervention studies against childhood obesity are being undertaken. Limited small intervention trials have shown some positive results. The results from the ongoing multi-center randomized nutrition and lifestyle intervention study by the China CDC [76] may provide some evidence for composing appropriate national strategies in relation to effective prevention and control of childhood obesity in China. Importantly, long-term effects of nutrition and lifestyle intervention on adulthood health outcomes merit carefully evaluation.

Collectively, several large-scale national surveys have released important health messages on the prevalence of and the time trends in childhood overweight and obesity in China. Most of the findings on childhood obesity and related metabolic disorders as well as the associated risk factors are based on cross-sectional analyses. It has been pointed out that there is currently no evidence-based clinically meaningful childhood obesity definition established [78]. Longitudinal studies may help estimate the extent to which excess weight during childhood may affect adulthood disease risk.

Some evidence has shown that multiple genetic markers are associated with greater infants' weight gain among Caucasians [79]. It remains unknown how genetic markers may influence weight gain among Chinese infants. The interactions between parents' genetic background, offspring's genetic variations, intrauterine development, and postnatal environment may have profound impact on childhood obesity development. In this regard, population-based birth cohort studies may provide valuable information for the identification of important factors related to excess weight gain during childhood. This may be particularly important for tailoring appropriate intervention strategies for Chinese children and adolescents in whom the prevalence of overweight and obesity is on increasing.

References

[1]   Yusuf S, Reddy S, Ounpuu S, Anand S. Global burden of cardiovascular diseases: part I: general considerations, the epidemiologic transition, risk factors, and impact of urbanization. *Circulation* 2001;104:2746-53.

[2]   He J, Gu D, Wu X, Reynolds K, Duan X, Yao C, et al. Major causes of death among men and women in China. *N Engl J Med* 2005;353:1124-34.

[3]   Lopez AD, Mathers CD, Ezzati M, Jamison DT, Murray CJ. Global and regional burden of disease and risk factors, 2001: systematic analysis of population health data. *Lancet* 2006;367:1747-57.

[4]   Willett WC, Dietz WH, Colditz GA. Guidelines for healthy weight. *N Engl J Med* 1999;341:427-34.

[5]   Gu D, Gupta A, Muntner P, Hu S, Duan X, Chen J, et al. Prevalence of cardiovascular disease risk factor clustering among the adult population of China: Results from the International Collaborative Study of Cardiovascular Disease in Asia (InterAsia). *Circulation* 2005;112:658-65.

[6]   Gu D, Reynolds K, Wu X, Chen J, Duan X, Reynolds RF, et al. Prevalence of the metabolic syndrome and overweight among adults in China. *Lancet* 2005;365:1398-405.

[7]   Wu Y. Overweight and obesity in China. *BMJ* 2006;333:362-3.

[8]   Yang W, Lu J, Weng J, Jia W, Ji L, Xiao J, et al. Prevalence of Diabetes among Men and Women in China. *N Engl J Med* 2010;362:1090-101.

[9]   Yang W, Reynolds K, Gu D, Chen J, He J. A comparison of two proposed definitions for metabolic syndrome in the Chinese adult population. *Am J Med Sci* 2007;334:184-9.

[10] Yu Z, Lin X, Haas JD, Franco OH, Rennie KL, Li H, et al. Obesity related metabolic abnormalities: Distribution and geographic differences among middle-aged and older Chinese populations. *Prev Med* 2009;48:272-8.

[11] Guo SS, Wu W, Chumlea WC, Roche AF. Predicting overweight and obesity in adulthood from body mass index values in childhood and adolescence. *Am J Clin Nutr* 2002;76:653-8.

[12] Vanhala M, Vanhala P, Kumpusalo E, Halonen P, Takala J. Relation between obesity from childhood to adulthood and the metabolic syndrome: population based study. *BMJ* 1998;317:319-20.

[13] Wang Y. Epidemiology of childhood obesity--methodological aspects and guidelines: what is new? *Int J Obes Relat Metab Disord* 2004;28 Suppl 3:S21-8.

[14] Ji CY. Report on childhood obesity in China (1)--body mass index reference for screening overweight and obesity in Chinese school-age children. *Biomed Environ Sci* 2005;18:390-400.

[15] Appropriate body-mass index for Asian populations and its implications for policy and intervention strategies. *Lancet* 2004;363:157-63.

[16] Ji CY, Cheng TO. Prevalence and geographic distribution of childhood obesity in China in 2005. *Int J Cardiol* 2008;131:1-8.

[17] Liu JM, Ye R, Li S, Ren A, Li Z, Liu Y. Prevalence of overweight/obesity in Chinese children. *Arch Med Res* 2007;38:882-6.

[18] Deckelbaum RJ, Williams CL. Childhood Obesity: The Health Issue. *Obesity* 2001;9:239S-43S.

[19] Wang Y. Cross-national comparison of childhood obesity: the epidemic and the relationship between obesity and socioeconomic status. *Int J Epidemiol* 2001;30:1129-36.

[20] Li Y, Schouten EG, Hu X, Cui Z, Luan D, Ma G. Obesity prevalence and time trend among youngsters in China, 1982-2002. *Asia Pac J Clin Nutr* 2008;17:131-7.

[21] Pan H, Jiang Y, Jing X, Fu S, Jiang Y, Lin Z, et al. Child body mass index in four cities of East China compared to Western references. *Ann Hum Biol* 2009;36:98-109.

[22] So HK, Nelson EA, Li AM, Wong EM, Lau JT, Guldan GS, et al. Secular changes in height, weight and body mass index in Hong Kong Children. *BMC Public Health* 2008;8:320.

[23] Zhou H, Yamauchi T, Natsuhara K, Yan Z, Lin H, Ichimaru N, et al. Overweight in urban schoolchildren assessed by body mass index and body fat mass in Dalian, China. *J Physiol Anthropol* 2006;25:41-8.

[24] Hui L, Bell AC. Overweight and obesity in children from Shenzhen, Peoples Republic of China. *Health & Place* 2003;9:371-6.

[25] Xu F, Li J, Ware RS, Owen N. Associations of television viewing time with excess body weight among urban and rural high-school students in regional mainland China. *Public Health Nutr* 2008;11:891-6.

[26] Shi Z, Lien N, Nirmal Kumar B, Dalen I, Holmboe-Ottesen G. The sociodemographic correlates of nutritional status of school adolescents in Jiangsu Province, China. *J Adolesc Health* 2005;37:313-22.

[27] Chunming C. Fat intake and nutritional status of children in China. *Am J Clin Nutr* 2000;72:1368S-72.

[28] Ji CY. The prevalence of childhood overweight/obesity and the epidemic changes in 1985-2000 for Chinese school-age children and adolescents. *Obesity Reviews* 2008;9:78-81.

[29] Popkin BM. Recent dynamics suggest selected countries catching up to US obesity. *Am J Clin Nutr* 2009:ajcn.2009.28473C.

[30] Luo J, Hu FB. Time trends of obesity in pre-school children in China from 1989 to 1997. *Int J Obes Relat Metab Disord* 2002;26:553-8.

[31] Wang Y, Monteiro C, Popkin BM. Trends of obesity and underweight in older children and adolescents in the United States, Brazil, China, and Russia. *Am J Clin Nutr* 2002;75:971-7.

[32] Liu J-M, Ye R, Li S, Ren A, Li Z, Liu Y, et al. Prevalence of Overweight/Obesity in Chinese Children. *Arch Med Res* 2007;38:882-6.

[33] Kong AP, Choi KC, Ko GT, Wong GW, Ozaki R, So WY, et al. Associations of overweight with insulin resistance, beta-cell function and inflammatory markers in Chinese adolescents. *Pediatr Diabetes* 2008;9:488-95.

[34] Zhang X, Liu E, Tian Z, Wang W, Ye T, Liu G, et al. High birth weight and overweight or obesity among Chinese children 3-6 years old. *Prev Med* 2009;49:172-8.

[35] Lear SA, Humphries KH, Kohli S, Chockalingam A, Frohlich JJ, Birmingham CL. Visceral adipose tissue accumulation differs according to ethnic background: results of the Multicultural Community Health Assessment Trial (M-CHAT). *Am J Clin Nutr* 2007;86:353-9.

[36] Wang Y, Mi J, Shan XY, Wang QJ, Ge KY. Is China facing an obesity epidemic and the consequences? The trends in obesity and chronic disease in China. *Int J Obes (Lond)* 2007;31:177-88.

[37] Li Y, Yang X, Zhai F, Piao J, Zhao W, Zhang J, et al. Childhood obesity and its health consequence in China. *Obes Rev* 2008;9 Suppl 1:82-6.

[38] Sung RYT, Tong PCY, Yu C-W, Lau PWC, Mok GTF, Yam M-C, et al. High Prevalence of Insulin Resistance and Metabolic Syndrome in Overweight/Obese Preadolescent Hong Kong Chinese Children Aged 9-12 Years. *Diabetes Care* 2003;26:250-1.

[39] Libby P, Ridker PM, Maseri A. Inflammation and atherosclerosis. *Circulation* 2002;105:1135-43.

[40] Pickup JC. Inflammation and Activated Innate Immunity in the Pathogenesis of Type 2 Diabetes. *Diabetes Care* 2004;27:813-23.

[41] Barker DJ. The fetal and infant origins of adult disease. *BMJ* 1990;301:1111.

[42] Boney CM, Verma A, Tucker R, Vohr BR. Metabolic Syndrome in Childhood: Association With Birth Weight, Maternal Obesity, and Gestational Diabetes Mellitus. *Pediatrics* 2005;115:e290-e6.

[43] Landon MB, Spong CY, Thom E, Carpenter MW, Ramin SM, Casey B, et al. A Multicenter, Randomized Trial of Treatment for Mild Gestational Diabetes. *N Engl J Med* 2009;361:1339-48.

[44] Whitaker RC, Pepe MS, Seidel KD, Wright JA, Knopp RH. Gestational diabetes and the risk of offspring obesity. *Pediatrics* 1998;101:E9.

[45] Yang X, Hsu-Hage B, Zhang H, Zhang C, Zhang Y. Women with impaired glucose tolerance during pregnancy have significantly poor pregnancy outcomes. *Diabetes Care* 2002;25:1619-24.

[46] Tam WH, Ma RC, Yang X, Ko GT, Tong PC, Cockram CS, et al. Glucose intolerance and cardiometabolic risk in children exposed to maternal gestational diabetes mellitus in utero. *Pediatrics* 2008;122:1229-34.

[47] Tam WH, Ma RCW, Yang X, Li AM, Ko GTC, Kong APS, et al. Glucose Intolerance and Cardiometabolic Risk in adolescent exposed to maternal Gestational Diabetes: a 15-year follow-up study. *Diabetes Care* 2010;33:1382-4

[48] Baker JL, Michaelsen KF, Rasmussen KM, Sorensen TI. Maternal prepregnant body mass index, duration of breastfeeding, and timing of complementary food introduction are associated with infant weight gain. *Am J Clin Nutr* 2004;80:1579-88.

[49] Gillman MW, Rifas-Shiman SL, Camargo CA, Jr., Berkey CS, Frazier AL, Rockett HR, et al. Risk of overweight among adolescents who were breastfed as infants. *JAMA* 2001;285:2461-7.

[50] Takahashi E, Yoshida K, Sugimori H, Miyakawa M, Izuno T, Yamagami T, et al. Influence factors on the development of obesity in 3-year-old children based on the Toyama study. *Prev Med* 1999;28:293-6.

[51] von Kries R, Koletzko B, Sauerwald T, von Mutius E, Barnert D, Grunert V, et al. Breast feeding and obesity: cross sectional study. *BMJ* 1999;319:147-50.

[52] Whitaker RC. Predicting preschooler obesity at birth: the role of maternal obesity in early pregnancy. *Pediatrics* 2004;114:e29-36.

[53] Parsons TJ, Power C, Manor O. Fetal and early life growth and body mass index from birth to early adulthood in 1958 British cohort: longitudinal study. *BMJ* 2001;323:1331-5.

[54] Wilson AC, Forsyth JS, Greene SA, Irvine L, Hau C, Howie PW. Relation of infant diet to childhood health: seven year follow up of cohort of children in Dundee infant feeding study. *BMJ* 1998;316:21-5.

[55] Yu Z, Sun JQ, Haas JD, Gu Y, Li Z, Lin X. Macrosomia is associated with high weight-for-height in children aged 1-3 years in Shanghai, China. *Int J Obes (Lond)* 2008;32:55-60.

[56] Popkin BM, Du S. Dynamics of the nutrition transition toward the animal foods sector in China and its implications: a worried perspective. *J Nutr* 2003;133:3898S-906S.

[57] Popkin BM, Keyou G, Zhai F, Guo X, Ma H, Zohoori N. The nutrition transition in China: a cross-sectional analysis. *Eur J Clin Nutr* 1993;47:333-46.

[58] Popkin BM. Comment: obesity patterns and the nutrition transition in China. *Arch Intern Med* 1994;154:2249, 53.

[59] Popkin BM, Horton S, Kim S, Mahal A, Shuigao J. Trends in diet, nutritional status, and diet-related noncommunicable diseases in China and India: the economic costs of the nutrition transition. *Nutr Rev* 2001;59:379-90.

[60] Du S, Mroz TA, Zhai F, Popkin BM. Rapid income growth adversely affects diet quality in China--particularly for the poor! *Soc Sci Med* 2004;59:1505-15.

[61] Wang Z, Zhai F, Du S, Popkin B. Dynamic shifts in Chinese eating behaviors. *Asia Pac J Clin Nutr* 2008;17:123-30.

[62] Liu Y, Zhai F, Popkin BM. Trends in eating behaviours among Chinese children (1991-1997). *Asia Pac J Clin Nutr* 2006;15:72-80.

[63] Li Y, Zhai F, Yang X, Schouten EG, Hu X, He Y, et al. Determinants of childhood overweight and obesity in China. *Br J Nutr* 2007;97:210-5.

[64] Tudor-Locke C, Ainsworth BE, Adair LS, Du S, Popkin BM. Physical activity and inactivity in Chinese school-aged youth: the China Health and Nutrition Survey. *Int J Obes Relat Metab Disord*;27:1093-9.

[65] Ma GS, Li YP, Hu XQ, Ma WJ, Wu J. Effect of television viewing on pediatric obesity. *Biomed Environ Sci* 2002;15:291-7.

[66] Cappuccio FP, Taggart FM, Kandala NB, Currie A, Peile E, Stranges S, et al. Meta-analysis of short sleep duration and obesity in children and adults. *Sleep* 2008;31:619-26.

[67] Chen X, Beydoun MA, Wang Y. Is Sleep Duration Associated With Childhood Obesity[quest] A Systematic Review and Meta-analysis. *Obesity* 2008;16:265-74.

[68] Spiegel K, Knutson K, Leproult R, Tasali E, Cauter EV. Sleep loss: a novel risk factor for insulin resistance and Type 2 diabetes. *J Appl Physiol* 2005;99:2008-19.

[69] Spiegel K, Leproult R, Van Cauter E. Impact of sleep debt on metabolic and endocrine function. *Lancet* 1999;354:1435-9.

[70] Yu Y, Lu BS, Wang B, Wang H, Yang J, Li Z, et al. Short sleep duration and adiposity in Chinese adolescents. *Sleep* 2007;30:1688-97.

[71] Tian Z, Ye T, Zhang X, Liu E, Wang W, Wang P, et al. Sleep Duration and Hyperglycemia Among Obese and Nonobese Children Aged 3 to 6 Years. *Arch Pediatr Adolesc Med*;164:46-52.

[72] Diabetes Prevention Program Research G. Reduction in the Incidence of Type 2 Diabetes with Lifestyle Intervention or Metformin. *N Engl J Med* 2002;346:393-403.

[73] Tuomilehto J, Lindstrom J, Eriksson JG, Valle TT, Hamalainen H, Ilanne-Parikka P, et al. Prevention of Type 2 Diabetes Mellitus by Changes in Lifestyle among Subjects with Impaired Glucose Tolerance. *N Engl J Med* 2001;344:1343-50.

[74] Jiang JX, Xia XL, Greiner T, Lian GL, Rosenqvist U. A two year family based behaviour treatment for obese children. *Arch Dis Child* 2005;90:1235-8.

[75] Hu C, Ye D, Li Y, Huang Y, Li L, Gao Y, et al. Evaluation of a kindergarten-based nutrition education intervention for pre-school children in China. *Public Health Nutr*;13:253-60.

[76] Li Y, Hu X, Zhang Q, Liu A, Fang H, Hao L, et al. The nutrition-based comprehensive intervention study on childhood obesity in China (NISCOC): a randomised cluster controlled trial. *BMC Public Health* 2010;10:229.

[77] Liu A, Hu X, Ma G, Cui Z, Pan Y, Chang S, et al. Evaluation of a classroom-based physical activity promoting programme. *Obes Rev* 2008;9 Suppl 1:130-4.

[78] Han JC, Lawlor DA, Kimm SY. Childhood obesity. *Lancet* 2010;375:1737-48.

[79] Elks CE, Loos RJ, Sharp SJ, Langenberg C, Ring SM, Timpson NJ, et al. Genetic markers of adult obesity risk are associated with greater early infancy weight gain and growth. *PLoS Med* 2010;7:e1000284.

In: Childhood Obesity: Risk Factors, Health Effects...
Editor: Carol M. Segel

ISBN: 978-1-61761-982-3
©2011 Nova Science Publishers, Inc.

Chapter IV

Ethnic Differences in Pediatric Obesity and the Metabolic Syndrome

Zeena Salman and Mark D. DeBoer
University of Virginia School of Medicine, VA, USA

Abstract

Childhood obesity is strongly linked to future diseases in later life, including cardiovascular disease (CVD) and Type 2 diabetes mellitus (T2DM). Prediction of these future diseases is assisted by diagnosis of the metabolic syndrome (MetS), a cluster of cardiovascular risk factors linked to insulin resistance. African Americans have higher rates of insulin resistance, T2DM and death from CVD but paradoxically have lower rates of MetS diagnosis. This is largely because of ethnic differences in triglyceride levels. If we are to accurately identify obese children who are in highest need of weight loss intervention, new risk predictors are needed that perform well among all ethnicities.

Introduction

Obesity in childhood has strong implications for increased morbidity and mortality, both during childhood and later in life (1-3). The increase in morbidity includes an increased risk of complications such as type 2 diabetes mellitus (T2DM) and cardiovascular disease (CVD) and thus threatens to shorten the lifespan of affected children (1, 4, 5). Since 1980, the prevalence of childhood overweight (defined as BMI-for-age 85th-95 percentile) and obesity (BMI-for-age $\geq$ 95th percentile) has more than doubled, and over the course of the past decade remain high at 15% and 17%, respectively (6-8). Also, the risk of an overweight/obese child becoming an overweight/obese adult is 79-89%, compared to the risk of a normal-weight child becoming an overweight adult of 11-22%, underscoring that the current childhood epidemic is likely to continue a worsening of obesity in adults (9, 10).

Early interventions involving weight loss through lifestyle modification have been found to be successful in decreasing the risk of these complications of obesity, but such interventions are both expensive as well as time-consuming, oftentimes requiring maintenance treatment for good long term outcomes (11-13). Because of the cost and effort involved in these treatments, it is important to choose a target population that would most benefit from such interventions. One such population is children diagnosed with metabolic syndrome (MetS), a cluster of cardiovascular factors associated with insulin resistance (Table 1(14, 15)). Children with MetS have a 12-fold increased risk of T2DM (16). Weight loss has been proven to produce resolution of MetS as well as to delay onset of T2DM (11, 17).

Table 1. Pediatric Metabolic Syndrome Criteria In the pediatric metabolic syndrome criteria adapted from the Adult Treatment Panel III definition in adults, individuals are diagnosed with MetS if they have ≥3 of findings listed among the individual components of MetS. In the International Diabetes Federation Criteria individuals are required to have an elevated waist circumference and two additional findings from the other components of MetS (Adapted from Ford et al. Circulation 2007; 115:2526-2532 and Zimmet et al. Lancet 2007; 369:2059-2061).

	Obesity	Hyper-tension	High TG (mg/dL)	HDL (mg/dL)	Elevated fasting glucose (mg/dL)
ATP III adaptation	WC ≥90%	SBP *or* DBP ≥90% for age, sex, height	TG ≥110 mg/dL	HDL ≤ 40	≥100
International Diabetes Federation	6-10 y.o.: WC ≥90%	≥130/85 mmHg	TG ≥150 mg/dL	HDL < 40	≥100
	10-16 y.o.: WC ≥90%	≥130/85 mmHg	TG ≥150 mg/dL	HDL: Males <40 Females < 50	≥100

Nevertheless, significant issues remain prior to using MetS as a decision point for increased intervention among all adolescents. African-American adolescents and adults have lower rates of MetS diagnosed, despite having higher rates to T2DM and CVD. In this article we will examine factors related to the low prevalence of MetS in African Americans and discuss potential alternate means of risk prediction for increased weight loss outcomes.

Ethnic Differences in T2DM and CVD

The main motivation for efforts at decreasing obesity in children and adolescents is to decrease future disease among affected children, the chief diseases being cardiovascular disease and T2DM. Both of these diseases have modifiable risk factors associated with them, in large part from the ability of individuals to improve their degree of insulin resistance as well as individual MetS components such as HDL cholesterol and blood pressure from lifestyle changes, including increased activity, improved dietary quality and overall weight loss. As we will see, both of these diseases have significant ethnic variation suggesting that not all of the population carries the same risk.

Cardiovascular disease continues to be the most common cause of death in the US (18). African Americans have had a longstanding history of higher death from CVD and stroke compared to white or Hispanic individuals (19). The reason for this higher death rate is unclear, though higher rates of hypertension among African Americans (discussed further below) and lower access to care have been cited (19).

African Americans also have a higher prevalence of T2DM. According to data from the National Health and Nutrition Survey '05-'06, 12.8% of non-Hispanic black individuals 20 y.o. or greater carried a diagnosis of diabetes, compared to 6.6% of non-Hispanic whites and 8.4% of Mexican Americans (20). This difference is true among adolescents as well, in that the incidence of T2DM among African American youth is several fold higher than among non-Hispanic whites and—at younger age groups—higher than among Hispanics (Table 2). While T2DM remains highly associated with obesity between all ethnic groups, as we will see, the high incidence of T2DM among African Americans belie differences between groups in obesity rates alone (21).

Table 2. Ethnic differences in Incidence of Type 2 Diabetes. Shown are incidence rates (and confidence intervals) of Diabetes (2002-2003) per 100 000 Person-Years by Age Group, Race/Ethnicity. Adapted from SEARCH for Diabetes Study Group, JAMA 2007;297(24):2716-2724.

	10-14 y.o.	15-19 y.o.
Non-Hispanic white	3.0 (2.3, 4.0)	5.6 (4.5, 6.9)
African American	22.3 (18.1, 27.5)	19.4 (15.3, 24.5)
Hispanic	8.9 (6.4, 12.3)	17.0 (13.3, 21.8)

Ethnic Differences in Insulin Resistance

Although impaired insulin secretion eventually becomes a factor in the production of T2DM, the predominant factor contributing to T2DM can be summed up in the term "insulin resistance" (22). This is a term that is used frequently but with surprisingly little clarity regarding its exact underlying mechanisms, which will be discussed further below (see "Ethnic differences in etiologies behind MetS,"). Insulin resistance refers to an elevated amount of insulin required to stimulate glucose uptake in a given individual. This is best studied using a euglycemic clamp study, where glucose is infused at a fixed rate and insulin is administered at the rate necessary to maintain stable, normal blood sugar readings. Individuals who have an element of insulin resistance require a large amount of infused insulin to maintain their blood sugar. A surrogate study to examine for insulin resistance is fasting insulin level. While obesity contributes to insulin resistance, the tendency toward insulin resistance frequently runs in families, suggesting a strong genetic basis (22).

African Americans have been consistently shown to have higher rates of insulin resistance than non-Hispanic whites. African American adolescents are approximately 20% less insulin sensitive than matched non-Hispanic whites, as shown by euglycemic clamp studies (23-25). During adolescence, African Americans also have higher fasting insulin levels despite a similar degree of adiposity (24, 26). The reason for these differences are not known entirely but are likely to include some genetic component, since these findings correlate with increasing African genetic admixture (27).

Ethnic Differences in MetS and its Components

Given ethnic differences in important parameters such as insulin resistance and outcomes such as T2DM and CVD, it would be reasonable to postulate that African Americans also have a higher rate of MetS and its individual components, each of which is associated with insulin resistance. In an unusual paradox, however, African Americans have a lower rate of MetS starting in childhood. Rates of MetS using the International Diabetes Federation criteria are shown in Figure 1, demonstrating that non-Hispanic black adolescent males have lower rates of MetS than do non-Hispanic whites and Mexican Americans (28). These differences are also present among adults, suggesting that longstanding factors contribute to this paradox (29).

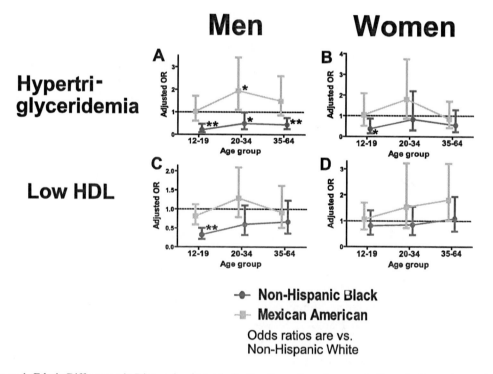

Figure 1. Ethnic Differences in Diagnosis of Metabolic Syndrome Prevalence and of metabolic syndrome by race/ethnicity for males (A) and females (B) across age groups using the pediatric and adult International Diabetes Federation definitions of MetS (NHANES 1999-2006). From Walker et al. Nutrition, Metabolism and Cardiovascular Disease, in press, used by permission.

In considering the roots these ethnic differences in MetS diagnosis, we will now consider the individual components used for MetS diagnosis. As described below, for most of the components, there are differences in both directions, with African Americans of one gender or the other sometimes exhibiting more severe elevations and sometimes exhibiting a lower tendency toward abnormalities in comparison to other ethnicities. There is one persistent difference between ethnicities, however: African Americans are consistently less likely to exhibit hypertriglyceridemia. As we will see, this is likely the source of the lower rates of MetS in comparison to other ethnicities.

Abdominal Obesity

As mentioned previously, MetS itself is comprised of multiple factors that cluster together more frequently than would be expected by chance. Although insulin resistance is considered the central feature of MetS, insulin resistance as described previously is not measured directly in the diagnosis of MetS. This is largely because of wide variability in the assays for insulin between different laboratories. Apart from insulin resistance, however, the next most central feature regarding MetS is felt to be *visceral adiposity*, most frequently assessed by measures of abdominal obesity. The prominence of visceral adiposity in the consideration of MetS is because of the tight connection between abdominal obesity and insulin resistance as seen in large scale studies (30-34). The importance of central obesity in diagnosing MetS is highlighted by making it a pre-requisite for diagnosis of MetS using the criteria of the International Diabetes Federation, a most-commonly used set of MetS criteria (15, 35).

In differentiating degrees of obesity between ethnicities, there are some complicating factors, in particular with respect to the measurement used to estimate abdominal obesity. The most accurate measurements include CT or MRI scans of the abdomen to assess the amount of visceral fat tissue. Clearly, these techniques are expensive and almost never warranted in assessing the adiposity of individual patients. The technique most frequently used is a standardized waist circumference measurement at the level of the umbilicus and the anterior iliac spine, a measurement that can be performed fairly easily in an outpatient clinic setting. Waist circumference measurement correlates well with visceral fat mass, though African Americans have a lower increase in visceral fat for each increment of waist circumference (36-38).

There is considerable ethnic variation in waist circumference ranges and in the threshold at which increasing WC confers increased insulin resistance. This is best seen among Asians, in which WC measurements in men become concerning for health risk at only 80 cm, compared to 94 cm for white individuals as given by the World Health Organization (35, 39-41). In addition to ethnicity-specific values for WC, in the case of children and adolescents, age needs to be taken into consideration. For this reason there are WC growth curves that can be used to estimate whether an adolescent's WC exceeds normal value for age (42).

Taking these ethnicity-specific and age-specific WC normal values into consideration, African American adolescent boys and men are less likely to have abdominal obesity than are whites and Hispanics (Mexican Americans). There is no difference in elevated WC in African American adolescent females as compared to whites and Mexican Americans, and among older ages, African American women 35-65 y.o. are more likely to exhibit an elevated WC (28, 29, 43).

As much as WC is the most commonly-used way to estimate abdominal obesity, it is not the most common way to diagnose obesity itself. In clinical settings, the most frequently-used measure to define obesity is body mass index (BMI). This is a measure of weight in kg divided by the square of the height in meters. As with WC, in children, there is a normal change in BMI with age, and thus BMI measurements always have to be assessed by percentile. Clinically, a BMI between the 85-95th percentile for age is defined as "overweight" and a BMI greater than or equal to the 95th percentile is defined as "obese."

Because BMI estimates overall weight and not simply abdominal obesity, it is not as good of a measure of future risk for insulin resistance. Individuals who have a high body weight because of higher amounts of muscle or subcutaneous fat may have a high BMI in the absence of significant visceral fat (44).

Ethnic differences in BMI start at an early age and persist through adulthood—particularly among African American girls. This is true even though there are not significant differences in WC among African American girls and women until later in adulthood (28). These higher BMI levels among African-American girls are at least in part due to higher rates of subcutaneous fat. In a study enrolling both African American and white girls and following them from age 9 to 19, it was noted that at the beginning and end of the study, the African American girls had higher BMI (by 0.4 kg/m2) and higher skinfold thickness (as a measure of total body fat).

Fasting Glucose

MetS criteria evaluate for elevated fasting glucose as a determinant of diabetes or pre-diabetes. Most MetS criteria—including the ATP III and IDF adult criteria and most pediatric adaptations of these criteria—utilize a fasting glucose level equal to or greater than 100 mg/dL as indicating elevated fasting glucose (14, 35), while a level above 125 mg/dL indicates frank diabetes. In most sets of pediatric MetS criteria, any fasting blood sugar above 100 mg/dL is sufficient to qualify as one of the positive findings required for diagnosis.

Ethnic differences in fasting blood sugar have been observed. Interestingly, although African American adolescents and adults are more likely to have T2DM, adolescent blacks are less likely to exhibit elevated fasting blood glucose values (9.4% vs. 14.8% for non-Hispanic whites and 15.2% for Hispanics)(43). This paradox may be explained by the etiology behind mildly-elevated fasting blood sugars (100-125 mg/dL) compared to T2DM (fasting blood sugar >125 mg/dL). Elevations in fasting blood sugars are felt to largely reflect hepatic insulin resistance, while diabetes is due to the combination of hepatic insulin resistance, peripheral insulin resistance and beta cell failure (45, 46). That African Americans have lower rates of elevated fasting blood sugar may indicate a lower rate of hepatic insulin resistance, though this is yet to be described further. Among adolescents, the lower rates of elevated fasting blood glucose likely reflect one of the reasons for lower prevalence of MetS diagnosis. As we will see, ethnic differences are more pronounced in other components of MetS.

Hypertension

Hypertension has long been noted to be more common among African Americans (45, 47). African American females have higher rates of hypertension across the age range starting by adolescence (28, 29, 48). Nevertheless, while African American adolescent boys and men tend toward higher blood pressures than other ethnicities, these differences fail to reach statistical until later years.(Johnson, Parks) Interestingly, Hispanic young men are less likely to have hypertension than whites (28). Overall, though, ethnic differences in blood pressure during adolescence are not large, with a prevalence of hypertension in non-Hispanic blacks of 9.1% vs. non-Hispanic whites 6.3% and Hispanics 6.9%. Thus, hypertension at this age is not likely to cause a significant difference in MetS diagnosis (43).

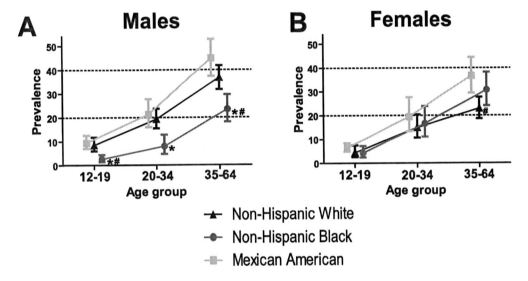

Figure 2. Ethnic Differences in Elevated Triglycerides and Low HDL. Odds Ratios for elevated triglyceride level and low HDL (per IDF MetS criteria) are shown for non-Hispanic blacks and Mexican Americans using non-Hispanic whites as the comparator group. OR's have been adjusted for education level, income:poverty ratio, physical inactivity and dietary quality (Healthy Eating Index). Data are from NHANES '99-'06. * $p<0.05$ vs. white non-Hispanic; ** $p<0.01$ vs. white non-Hispanics. From Walker et al. Nutrition, Metabolism and Cardiovascular Disease, in press, used by permission.

Low HDL-Cholesterol

Because HDL-C serves as a scavenger of excess cholesterol in circulation, low levels of HDL-C confer an increase risk for atherosclerosis. HDL-C is removed from circulation in part by hepatic lipase, an enzyme whose activity is increased in the setting of insulin resistance (49). There are significant ethnic differences in the levels of HDL-C, starting in adolescence. Non-Hispanic black adolescent males have a much lower prevalence of low HDL-C than seen in other groups (non-Hispanic blacks 10.6%, non-Hispanic whites 27.9%, Hispanics 24.0%), and this continues through the lifespan (Figure 2). This trend is also present among African American women, though the differences are less severe and in many cases fail to reach statistical significance. These differences are interesting given the higher rates of insulin resistance among African Americans. The normal levels of HDL despite insulin resistance may be explained by lower levels of hepatic lipase among African Americans, even in the setting of insulin resistance (50, 51). Whatever the reason for these findings, the low rates of low HDL among African Americans clearly do not appear to confer significant cardioprotection and are likely a source of the lower rates of MetS diagnosis among African Americans.

Triglycerides

The widest ethnic difference among all MetS components are seen regarding triglyceride levels. In adolescence, elevated triglyceride levels ($\geq$110 mg/dL) are seen in only 9.1% of non-Hispanic blacks, compared to 28.9% of non-Hispanic whites and 26.1% of Hispanics (43). These extreme differences are seen among males and females and exist across the age spectrum (29, 43). The reasons for this are analogous to what is seen for HDL-C and hepatic lipase, mentioned above. In this case triglycerides are normally removed from circulation via the action of lipoprotein lipase. Lipoprotein lipase in turn has decreased activity in the setting of insulin resistance, leading to higher levels of triglycerides. Among African Americans, the action of lipoprotein lipase is higher than in other ethnicities (52-54). It is notable that increases in triglyceride levels are correlated with insulin resistance among non-Hispanic blacks, non-Hispanic whites and Mexican Americans (55). This suggests that non-Hispanic blacks start with a lower baseline of triglycerides, and though these triglyceride levels increase with worsening degrees of insulin resistance, they are still unlikely to reach the cut-off set as part of MetS criteria. All told, the difference in triglyceride levels between ethnicities is the major reason for ethnic differences in MetS.

Summary of Ethnic Differences in MetS Components

Thus, in considering the components of MetS, African American adolescents have similar rates of elevated WC and hypertension, a slightly lower prevalence of elevated blood glucose, and have a significantly lower prevalence of hypertriglyceridemia and low HDL-C. Given a rate of hypertriglyceridemia that is less than a third of what is seen in other ethnicities, the difference in triglyceride levels are the most likely reason that African Americans are diagnosed with MetS at such lower rates. Whether African Americans actually experience less of the underlying pathophysiology behind MetS is more up for debate and will be considered in the following section.

Ethnic Differences in Etiologies Behind MetS

In considering the cellular and systemic processes responsible for producing insulin resistance and the components of MetS, it is apparent that there are multiple overlapping events that contribute to these clinical findings. Insulin resistance includes issues as seemingly diverse as systemic inflammation, production of hormones by fat tissue (referred to as "adipokines") and—ultimately—a decrease in the effectiveness of insulin to stimulate glucose uptake (56, 57).

As is suggested by its prominence in the IDF MetS criteria (which uses waist circumference as a necessary criterion), visceral adipose tissue is felt to play a central role in the processes underlying MetS. This is due to dysfunction of hypertrophied visceral adipocytes, including a decrease in secretion of a key adipokine called adiponectin. Normally adiponectin suppresses inflammation by decreasing the expression and action of inflammatory cytokines (58, 59). Thus, when adiponectin secretion decreases as a result of visceral obesity, this produces increased inflammation—among other effects—that results in a worsening of insulin resistance (60-63).

Several experimental observations suggest that adiponectin is not merely a marker of worsening insulin resistance but plays a primary role. Adiponectin levels increase during weight loss, coincident with decreasing inflammation (64, 65). Even more convincingly, though, when adiponectin was administered to laboratory animals, this causes an improvement in insulin sensitivity and inflammation markers (59, 66).

In addition to the effects adiponectin has on inflammation, there are other causes of increased inflammation in the setting of obesity. With worsening central adiposity, visceral adipocytes release increasing amounts of both inflammatory cytokines as well as chemoattractants that recruit macrophages to the area (57, 67-69). Ultimately, systemic inflammation promotes phosphorylation of the insulin receptor, contributing to insulin resistance (22).

Although there are likely a multitude of other processes that are involved in producing the findings of MetS, in considering ethnic differences we will focus on low levels of adiponectin and increased inflammation due to their strong link with insulin resistance. As suggested previously, the lower prevalence of MetS among African Americans could be construed as indicating that these processes underlying MetS are also less common. As we will see, that does not appear to be the case.

In the case of adiponectin, African American adolescents have been shown to have lower levels than seen in whites, even when considering children of similar age, body habitus and pubertal stage(Table 3 (25)). These lower levels of adiponectin among African Americans are more consistent with the higher rates of insulin resistance than the lower rates of MetS diagnosis. Given that adiponectin appears to be of primary importance in producing insulin resistance, levels of adiponectin in African Americans may represent a more accurate estimate of future risk than is portrayed by a diagnosis of MetS.

The same is true for markers of inflammation. The most commonly-used marker of systemic inflammation is high-sensitivity C-reactive protein (hs-CRP). In large prospective studies, high levels of hsCRP have been shown to precede development of cardiovascular disease and T2DM. African Americans have been shown to have either similar levels (70) or increased levels of hsCRP (71-73), but not lower levels as may be suspected by their lower rates of MetS diagnosis. In adolescents, non-Hispanic blacks have higher levels of hsCRP than seen among non-Hispanic whites but similar levels to that seen in Mexican Americans (74). Non-Hispanic black adolescents with MetS have much higher levels of hsCRP than seen in non-Hispanic white and Mexican American adolescents with MetS, suggesting that by the time African Americans are diagnosed with MetS by currently-used criteria, they exhibit a more advanced condition (Table 4(74)). Overall, the high rates of elevated hsCRP in African Americans again suggest that current-used MetS criteria are not adequately capturing cardiovascular and diabetes risk among African American individuals.

Table 3. Ethnic differences in levels of adiponectin and insulin sensitivity in adolescents. Adiponectin levels and insulin sensitivity are lower in African American adolescents compared to whites who are well-matched in age, gender and BMI. Data obtained during euglycermic clamp study in which glucose was infused at a fixed rate in each individual and insulin was infused the rate necessary to maintain constant blood sugar levels. Adapted from Lee Diabetes Care 2006; 29:51-56.

	African American	Caucasian
N (male/female)	41/42	41/37
Age (range)	12.2 (8.5-16.8)	12.3 (8.0-17.2)
BMI (kg/m2, mean +/-SEM)	24.8 +/- 0.9	24.6 +/- 0.96
Adiponectin (ug/mL, mean +/-SEM)	10.2 +/- 0.6 *	12.1 +/- 0.68
Insulin sensitivity (mgxkg /FFMxmin, mean +/-SEM)	11.0 +/- 0.65 *	13.2 +/- 0.91

Table 4. Ethnic Differences in Levels of hsCRP Metabolic Syndrome Status in Adolescents. Data are from NHANES 1999-2006, using an adaptation of the ATP III criteria for MetS. From DeBoer et al. Endocrine Reviews 2010; 31:S1410.

	NHB n=886	NHW n=778	MA n=1018	P-value[+]
Percent with MetS (%)	4.3	8.9	9.6	a**, b**
hsCRP mg/dL (mean±SE)	1.22 (0.05)	0.95 (0.06)	1.21 (0.06)	a**, c**
hsCRP with MetS	3.09 (0.45)	1.73 (0.16)	2.00 (0.17)	a**, b*
hsCRP without MetS	1.06 (0.06)	0.89 (0.06)	1.14 (0.06)	a**, c**
Difference in hsCRP, MetS(+) – MetS (-)	2.02 (0.48)	0.84 (0.17)	0.86 (0.17)	a*, b*

[a] NHB vs. NHW; [b] NHB vs. MA; [c] NHW vs. MA; *<0.05, **<0.01

Potential Alternatives for Risk Assessment

Need for Predictive Factors

What are direly needed in the care of obese children are factors that are present in childhood and accurately predict progression to adult disease. Armed with such tools, providers would be able to focus particularly strong attention on weight loss efforts for subjects at highest risk for future disease. It may also be true that children and families found to be at higher risk would be more motivated toward lifestyle changes that could improve their risk status.

Despite this strong need, it should be noted that it is a difficult process to identify childhood predictive factors, since such studies require longitudinally-collected data over a span long enough to include adult disease outcomes. This usually means following a cohort over a period of 20 years or more. While there is a growing list of these prospective studies beginning in childhood, little data is available at this point regarding future risk. What is also important to the subject matter at hand is that many of prospective studies published thus far have involved groups of children that were almost exclusively white.

Interestingly, the childhood predictor that has been most robust in its ability to predict future T2DM was a diagnosis of MetS, with an odds ratio of 11.5 for predicting diabetes 25-30 years later (16). It is unclear how MetS would perform as a predictive tool among African Americans, however. As discussed above, despite having low rates of MetS diagnosis, African Americans have high rates of obesity, insulin resistance, cardiovascular disease and T2DM. Thus, it would be expected that using a MetS diagnosis as a predictor of future disease would fail to identify a large proportion of African American adolescents at high risk. Alternate means of accurately predicting future risk among African Americans are likely to be needed.

Alternate MetS Criteria

One means of more accurate risk prediction across ethnicities could be via formulation of ethnicity-specific MetS diagnostic criteria. These criteria could be achieved by use of ethnicity-specific cut points to define elevations in MetS components. For example, given that African Americans have lower triglyceride levels, a lower level of triglycerides could be used to identify individuals who have triglyceride levels that are elevated compared to normal values among African Americans, even if these do not meet the triglyceride cut-off for the entire population (55). On this basis, it would be expected that more African Americans would meet criteria for a diagnosis of MetS. This could potentially improve the ability of MetS to predict future disease.

Another means that has been proposed by many is a linear MetS score that quantitates MetS risk based on the extent of elevated values of multiple components, instead of requiring three or more components to be elevated above a specific cut-point (75-77). Individuals who had modest elevations in all 5 components (even if none of these values exceeded the current cut-points) may have a similar linear MetS score as individuals with extreme elevations in only 3 of the components. This type of system may better predict risk in African Americans who have significantly elevated levels in waist circumference and fasting glucose, even if they do not have high enough levels of triglycerides and low enough levels of HDL to currently trigger MetS diagnosis.

Family History

One marker of risk of risk that is present in childhood is the presence of a family history of T2DM. It has long been known that risk for T2DM has significant hereditary components to it and this appears to be true among multiple different ethnicities, including African Americans (78). One prospective study following children who later when on to develop T2DM found an odds ratio of 5.4 for future T2DM among those whose parents had T2DM compared to those with parents without T2DM (79). The predictive ability of more distant family members having diabetes is not known. However, given these risks and the strong hereditary nature of diabetes, it is reasonable that physicians place a strong emphasis on family history in targeting children for risk. Nevertheless, most children who go on to develop T2DM will not have a positive family history, and other markers are likely to be beneficial.

Serum Markers

Similar to the serum markers that make up MetS components, there is great interest in discovering additional markers that could be harbingers of greater risk for adult disease. One of these that is frequently used clinically is insulin. High levels of insulin in some ethnic populations have been shown to predict future T2DM (80, 81). Unfortunately, as mentioned previously, insulin testing is not standardized, and thus there are not established cut-offs to indicate what level constitutes increased risk. Nevertheless, insulin may still have a future in the prediction of future risk for diabetes.

Oral glucose tolerance tests can detect in a more sensitive nature some of the early changes in insulin sercretion and action that precede the development of T2DM. In a study of 117 obese children seen at an obesity treatment clinic 24% of those with an abnormal glucose tolerance test went on to T2DM within an average of 2 years (82).

Finally, new markers not currently used clinically among adolescents are needed. One of the possible markers is adiponectin, the adipokine that is inversely associated with insulin sensitivity (Table 3). As mentioned previously, adiponectin levels are lower (i.e., more concerning) among African American adolescents than among white adolescents, paralleling the increased risk of T2DM and CVD seen among African Americans. This may avoid the ethnic bias of a diagnosis of MetS. Adiponectin showed some promise in predicting disease progression in one longitudinal study in which low levels of adiponectin predicted worsening of MetS components 6 years later (83). Unfortunately, this study was performed in Denmark and did not include children of African descent. Further studies with more diverse ethnic representation are needed, as are studies with a much longer follow-up period.

In adults hsCRP has proven to be a reasonable marker of risk in that compared to those with low hsCRP values, adult women with an hsCRP value greater than 2.0 mg/L are more than twice as likely to experience a serious cardiovascular event as those with a hsCRP less than 0.49 mg/L (84) and those with an hsCRP level greater than 2.7 were more than four times as likely to develop T2DM than those less than 0.5 mg/L (85). As mentioned previously, hsCRP does not appear to have the same ethnic bias as MetS does, though data remain lacking regarding its use among different ethnicities.

Emphasizing the status of hsCRP as a marker of risk, the FDA has approved elevated hsCRP as a criteria for statin treatment in older adults without LDL cholesterol (statins are known to decrease systemic inflammation in addition to their lipid lowering properties). Unfortunately, no prospective data regarding hsCRP is available from childhood, and there is some reason to believe that the frequent intercurrent illness seen in childhood may make this a less accurate predictor of future disease.

The serum markers discussed here—or others—will be tested in prospective trials and will be able to improve our current ability to predict future disease risk on an ethnically-equal basis. It is hoped—but also remains to be proven—that more accurate risk prediction will assist in targeting patients that our in highest need for weight loss interventions. Clearly, these efforts will need to be sensitive to the ethnic differences that we have discussed, such that no ethnic group inappropriately misses out on such increased intervention efforts.

Conclusion

In conclusion, childhood obesity is associated with risk for future diseases including CVD and T2DM, though we currently lack tools to more accurately predict this risk among given children. Though the adult diseases associated with obesity are more prevalent among African Americans, our current risk predictors—such as MetS—are not equally effective at identifying risks among African Americans. If we are to accurately identify obese children who are in highest need of weight loss intervention, new risk predictors are needed that perform well among all ethnicities.

References

[1]   Franks PW, Hanson RL, Knowler WC, Sievers ML, Bennett PH, Looker HC. Childhood obesity, other cardiovascular risk factors, and premature death. *N Engl J Med* 2010;362(6):485-93.

[2]   Freedman DS, Dietz WH, Srinivasan SR, Berenson GS. The relation of overweight to cardiovascular risk factors among children and adolescents: the Bogalusa Heart Study. *Pediatrics* 1999;103(6 Pt 1):1175-82.

[3]   Must A, Spadano J, Coakley EH, Field AE, Colditz G, Dietz WH. The disease burden associated with overweight and obesity. *Jama* 1999;282(16):1523-9.

[4]   Narayan KM, Boyle JP, Thompson TJ, Sorensen SW, Williamson DF. Lifetime risk for diabetes mellitus in the United States. *Jama* 2003;290(14):1884-90.

[5]   Olshansky SJ, Passaro DJ, Hershow RC, Layden J, Carnes BA, Brody J, et al. A potential decline in life expectancy in the United States in the 21st century. *N Engl J Med* 2005;352(11):1138-45.

[6]   Ogden CL, Carroll MD, Curtin LR, Lamb MM, Flegal KM. Prevalence of high body mass index in US children and adolescents, 2007-2008. *Jama* 2010;303(3):242-9.

[7]   Ogden CL, Carroll MD, Curtin LR, McDowell MA, Tabak CJ, Flegal KM. Prevalence of overweight and obesity in the United States, 1999-2004. *Jama* 2006;295(13):1549-55.

[8]   Ogden CL, Carroll MD, Flegal KM. High body mass index for age among US children and adolescents, 2003-2006. *Jama* 2008;299(20):2401-5.

[9]   Freedman DS, Dietz WH, Srinivasan SR, Berenson GS. Risk factors and adult body mass index among overweight children: the Bogalusa Heart Study. *Pediatrics* 2009;123(3):750-7.

[10]   Stovitz SD, Hannan PJ, Lytle LA, Demerath EW, Pereira MA, Himes JH. Child height and the risk of young-adult obesity. *Am J Prev Med* 2010;38(1):74-7.

[11]   Knowler WC, Barrett-Connor E, Fowler SE, Hamman RF, Lachin JM, Walker EA, et al. Reduction in the incidence of type 2 diabetes with lifestyle intervention or metformin. *N Engl J Med* 2002;346(6):393-403.

[12]   Savoye M, Shaw M, Dziura J, Tamborlane WV, Rose P, Guandalini C, et al. Effects of a weight management program on body composition and metabolic parameters in overweight children: a randomized controlled trial. *Jama* 2007;297(24):2697-704.

[13] Wilfley DE, Stein RI, Saelens BE, Mockus DS, Matt GE, Hayden-Wade HA, et al. Efficacy of maintenance treatment approaches for childhood overweight: a randomized controlled trial. *Jama* 2007;298(14):1661-73.

[14] Ford ES, Li C, Cook S, Choi HK. Serum concentrations of uric acid and the metabolic syndrome among US children and adolescents. *Circulation* 2007;115(19):2526-32.

[15] Zimmet P, Alberti G, Kaufman F, Tajima N, Silink M, Arslanian S, et al. The metabolic syndrome in children and adolescents. *Lancet* 2007;369(9579):2059-61.

[16] Morrison JA, Friedman LA, Wang P, Glueck CJ. Metabolic syndrome in childhood predicts adult metabolic syndrome and type 2 diabetes mellitus 25 to 30 years later. *J Pediatr* 2008;152(2):201-6.

[17] Coppen AM, Risser JA, Vash PD. Metabolic syndrome resolution in children and adolescents after 10 weeks of weight loss. *J Cardiometab Syndr* 2008;3(4):205-10.

[18] CDC. Deaths: Leading Causes for 2006. *National Vital Statistics Reports* 2010; 58(14):1-100.

[19] Mensah GA, Mokdad AH, Ford ES, Greenlund KJ, Croft JB. State of disparities in cardiovascular health in the United States. *Circulation* 2005;111(10):1233-41.

[20] Cowie CC, Rust KF, Byrd-Holt DD, Gregg EW, Ford ES, Geiss LS, et al. Prevalence of diabetes and high risk for diabetes using A1C criteria in the U.S. population in 1988-2006. *Diabetes Care* 2010;33(3):562-8.

[21] Dabelea D, Bell RA, D'Agostino RB, Jr., Imperatore G, Johansen JM, Linder B, et al. Incidence of diabetes in youth in the United States. *Jama* 2007;297(24):2716-24.

[22] Defronzo RA. Banting Lecture. From the triumvirate to the ominous octet: a new paradigm for the treatment of type 2 diabetes mellitus. *Diabetes* 2009;58(4):773-95.

[23] Arslanian S, Suprasongsin C. Differences in the in vivo insulin secretion and sensitivity of healthy black versus white adolescents. *J Pediatr* 1996;129(3):440-3.

[24] Arslanian S, Suprasongsin C, Janosky JE. Insulin secretion and sensitivity in black versus white prepubertal healthy children. *J Clin Endocrinol Metab* 1997;82(6):1923-7.

[25] Lee S, Bacha F, Gungor N, Arslanian SA. Racial differences in adiponectin in youth: relationship to visceral fat and insulin sensitivity. *Diabetes Care* 2006;29(1):51-6.

[26] Messiah SE, Arheart KL, Lipshultz SE, Miller TL. Body mass index, waist circumference, and cardiovascular risk factors in adolescents. *J Pediatr* 2008;153(6):845-50.

[27] Casazza K, Phadke RP, Fernandez JR, Watanabe RM, Goran MI, Gower BA. Obesity attenuates the contribution of African admixture to the insulin secretory profile in peripubertal children: a longitudinal analysis. *Obesity* (Silver Spring) 2009;17(7):1318-25.

[28] Walker SE, Gurka MJ, Oliver MN, Johns D, DeBoer MD. Racial/ethnic discrepancies in the metabolic syndrome begin in childhood and persist after adjustment for environmental factors. *Nutr Metab Cardiovasc Dis* 2010;In press.

[29] Park YW, Zhu S, Palaniappan L, Heshka S, Carnethon MR, Heymsfield SB. The metabolic syndrome: prevalence and associated risk factor findings in the US population from the Third National Health and Nutrition Examination Survey, 1988-1994. *Arch Intern Med* 2003;163(4):427-36.

[30] Lee S, Bacha F, Gungor N, Arslanian SA. Waist circumference is an independent predictor of insulin resistance in black and white youths. *J Pediatr* 2006;148(2):188-94.

[31] Despres JP, Nadeau A, Tremblay A, Ferland M, Moorjani S, Lupien PJ, et al. Role of deep abdominal fat in the association between regional adipose tissue distribution and glucose tolerance in obese women. *Diabetes* 1989;38(3):304-9.

[32] Ross R, Fortier L, Hudson R. Separate associations between visceral and subcutaneous adipose tissue distribution, insulin and glucose levels in obese women. *Diabetes Care* 1996;19(12):1404-11.

[33] Ross R, Aru J, Freeman J, Hudson R, Janssen I. Abdominal adiposity and insulin resistance in obese men. *Am J Physiol Endocrinol Metab* 2002;282(3):E657-63.

[34] Tulloch-Reid MK, Hanson RL, Sebring NG, Reynolds JC, Premkumar A, Genovese DJ, et al. Both subcutaneous and visceral adipose tissue correlate highly with insulin resistance in african americans. *Obes Res* 2004;12(8):1352-9.

[35] Alberti KG, Zimmet P, Shaw J. Metabolic syndrome--a new world-wide definition. A Consensus Statement from the International Diabetes Federation. *Diabet Med* 2006;23(5):469-80.

[36] Carroll JF, Chiapa AL, Rodriquez M, Phelps DR, Cardarelli KM, Vishwanatha JK, et al. Visceral fat, waist circumference, and BMI: impact of race/ethnicity. *Obesity* (Silver Spring) 2008;16(3):600-7.

[37] Conway JM, Yanovski SZ, Avila NA, Hubbard VS. Visceral adipose tissue differences in black and white women. *Am J Clin Nutr* 1995;61(4):765-71.

[38] Perry AC, Applegate EB, Jackson ML, Deprima S, Goldberg RB, Ross R, et al. Racial differences in visceral adipose tissue but not anthropometric markers of health-related variables. *J Appl Physiol* 2000;89(2):636-43.

[39] Cameron AJ, Sicree RA, Zimmet PZ, Alberti KGT, A.M., Balkau B, Tuomilehto J, et al. Cut-points for Waist Circumference in Europids and South Asians. *Obesity* (Silver Spring) 2010:e-pub ahead of print.

[40] Vikram NK, Pandey RM, Misra A, Sharma R, Devi JR, Khanna N. Non-obese (body mass index < 25 kg/m2) Asian Indians with normal waist circumference have high cardiovascular risk. *Nutrition* 2003;19(6):503-9.

[41] Wang F, Wu S, Song Y, Tang X, Marshall R, Liang M, et al. Waist circumference, body mass index and waist to hip ratio for prediction of the metabolic syndrome in Chinese. *Nutr Metab Cardiovasc Dis* 2009;19(8):542-7.

[42] Li C, Ford ES, Mokdad AH, Cook S. Recent trends in waist circumference and waist-height ratio among US children and adolescents. *Pediatrics* 2006;118(5):e1390-8.

[43] Johnson WD, Kroon JJ, Greenway FL, Bouchard C, Ryan D, Katzmarzyk PT. Prevalence of risk factors for metabolic syndrome in adolescents: National Health and Nutrition Examination Survey (NHANES), 2001-2006. *Arch Pediatr Adolesc Med* 2009;163(4):371-7.

[44] Hull HR, Thornton J, Wang J, Pierson RNJ, Kaleem Z, Pi-Sunyer X, et al. Fat-free mass index: changes and race/ethnic differences in adulthood. *Int J Obes* (Lond). 2010:e-pub ahead of print.

[45] Sumner AE. Ethnic Differences in Triglyceride Levels and High-Density Lipoprotein Lead to Underdiagnosis of the Metabolic Syndrome in Black Children and Adults. *Journal of Pediatrics* 2009;155(S7):e7-e11.

[46] Nathan DM, Davidson MB, DeFronzo RA, Heine RJ, Henry RR, Pratley R, et al. Impaired fasting glucose and impaired glucose tolerance: implications for care. *Diabetes Care* 2007;30(3):753-9.

[47] Winkleby MA, Robinson TN, Sundquist J, Kraemer HC. Ethnic variation in cardiovascular disease risk factors among children and young adults: findings from the Third National Health and Nutrition Examination Survey, 1988-1994. *Jama* 1999;281(11):1006-13.

[48] Cook S, Auinger P, Li C, Ford ES. Metabolic syndrome rates in United States adolescents, from the National Health and Nutrition Examination Survey, 1999-2002. *J Pediatr* 2008;152(2):165-70.

[49] Blades B, Vega GL, Grundy SM. Activities of lipoprotein lipase and hepatic triglyceride lipase in postheparin plasma of patients with low concentrations of HDL cholesterol. *Arterioscler Thromb* 1993;13(8):1227-35.

[50] Clarenbach JJ, Vega GL, Adams-Huet B, Considine RV, Ricks M, Sumner AE. Variability in postheparin hepatic lipase activity is associated with plasma adiponectin levels in African Americans. *J Investig Med* 2007;55(4):187-94.

[51] Vega GL, Clark LT, Tang A, Marcovina S, Grundy SM, Cohen JC. Hepatic lipase activity is lower in African American men than in white American men: effects of 5' flanking polymorphism in the hepatic lipase gene (LIPC). *J Lipid Res* 1998;39(1):228-32.

[52] Bower JF, Deshaies Y, Pfeifer M, Tanenberg RJ, Barakat HA. Ethnic differences in postprandial triglyceride response to a fatty meal and lipoprotein lipase in lean and obese African American and Caucasian women. *Metabolism* 2002;51(2):211-7.

[53] Despres JP, Couillard C, Gagnon J, Bergeron J, Leon AS, Rao DC, et al. Race, visceral adipose tissue, plasma lipids, and lipoprotein lipase activity in men and women: the Health, Risk Factors, Exercise Training, and Genetics (HERITAGE) family study. *Arterioscler Thromb Vasc Biol* 2000;20(8):1932-8.

[54] Friday KE, Srinivasan SR, Elkasabany A, Dong C, Wattigney WA, Dalferes E, Jr., et al. Black-white differences in postprandial triglyceride response and postheparin lipoprotein lipase and hepatic triglyceride lipase among young men. *Metabolism* 1999;48(6):749-54.

[55] Sumner AE, Cowie CC. Ethnic differences in the ability of triglyceride levels to identify insulin resistance. *Atherosclerosis* 2008;196(2):696-703.

[56] de Ferranti S, Mozaffarian D. The perfect storm: obesity, adipocyte dysfunction, and metabolic consequences. *Clin Chem* 2008;54(6):945-55.

[57] Tilg H, Moschen AR. Inflammatory mechanisms in the regulation of insulin resistance. *Mol Med* 2008;14(3-4):222-31.

[58] Berg AH, Combs TP, Scherer PE. ACRP30/adiponectin: an adipokine regulating glucose and lipid metabolism. *Trends Endocrinol Metab* 2002;13(2):84-9.

[59] Maeda N, Shimomura I, Kishida K, Nishizawa H, Matsuda M, Nagaretani H, et al. Diet-induced insulin resistance in mice lacking adiponectin/ACRP30. *Nat Med* 2002;8(7):731-7.

[60] Huang KC, Lue BH, Yen RF, Shen CG, Ho SR, Tai TY, et al. Plasma adiponectin levels and metabolic factors in nondiabetic adolescents. *Obes Res* 2004;12(1):119-24.

[61] Ouchi N, Kihara S, Arita Y, Okamoto Y, Maeda K, Kuriyama H, et al. Adiponectin, an adipocyte-derived plasma protein, inhibits endothelial NF-kappaB signaling through a cAMP-dependent pathway. *Circulation* 2000;102(11):1296-301.

[62] Stefan N, Bunt JC, Salbe AD, Funahashi T, Matsuzawa Y, Tataranni PA. Plasma adiponectin concentrations in children: relationships with obesity and insulinemia. *J Clin Endocrinol Metab* 2002;87(10):4652-6.

[63] Weyer C, Funahashi T, Tanaka S, Hotta K, Matsuzawa Y, Pratley RE, et al. Hypoadiponectinemia in obesity and type 2 diabetes: close association with insulin resistance and hyperinsulinemia. *J Clin Endocrinol Metab* 2001;86(5):1930-5.

[64] Lazzer S, Vermorel M, Montaurier C, Meyer M, Boirie Y. Changes in adipocyte hormones and lipid oxidation associated with weight loss and regain in severely obese adolescents. *Int J Obes* (Lond) 2005;29(10):1184-91.

[65] Roth CL, Kratz M, Ralston MM, Reinehr T. Changes in adipose-derived inflammatory cytokines and chemokines after successful lifestyle intervention in obese children. *Metabolism* 2010:e-pub ahead of print.

[66] Yamauchi T, Kamon J, Waki H, Terauchi Y, Kubota N, Hara K, et al. The fat-derived hormone adiponectin reverses insulin resistance associated with both lipoatrophy and obesity. *Nat Med* 2001;7(8):941-6.

[67] Martin LJ, Woo JG, Daniels SR, Goodman E, Dolan LM. The relationships of adiponectin with insulin and lipids are strengthened with increasing adiposity. *J Clin Endocrinol Metab* 2005;90(7):4255-9.

[68] Aguirre V, Uchida T, Yenush L, Davis R, White MF. The c-Jun NH(2)-terminal kinase promotes insulin resistance during association with insulin receptor substrate-1 and phosphorylation of Ser(307). *J Biol Chem* 2000;275(12):9047-54.

[69] Harman-Boehm I, Bluher M, Redel H, Sion-Vardy N, Ovadia S, Avinoach E, et al. Macrophage infiltration into omental versus subcutaneous fat across different populations: effect of regional adiposity and the comorbidities of obesity. *J Clin Endocrinol Metab* 2007;92(6):2240-7.

[70] Ford ES, Giles WH, Myers GL, Rifai N, Ridker PM, Mannino DM. C-reactive protein concentration distribution among US children and young adults: findings from the National Health and Nutrition Examination Survey, 1999-2000. *Clin Chem* 2003;49(8):1353-7.

[71] Albert MA, Glynn RJ, Buring J, Ridker PM. C-reactive protein levels among women of various ethnic groups living in the United States (from the Women's Health Study). *Am J Cardiol* 2004;93(10):1238-42.

[72] Khera A, McGuire DK, Murphy SA, Stanek HG, Das SR, Vongpatanasin W, et al. Race and gender differences in C-reactive protein levels. *J Am Coll Cardiol* 2005;46(3):464-9.

[73] Wee CC, Mukamal KJ, Huang A, Davis RB, McCarthy EP, Mittleman MA. Obesity and C-reactive protein levels among white, black, and hispanic US adults. *Obesity* (Silver Spring) 2008;16(4):875-80.

[74] DeBoer MD, Gurka MJ, Sumner AE. Inflammation is Worse in Non-Hispanic Black Adolescents with Metabolic Syndrome than Non-Hispanic Whites or Mexican Americans: NHANES 1999-2006. *Endocrine Reviews* 2010;31(3):S1410.

[75] Thivel D, Malina RM, Isacco L, Aucouturier J, Meyer M, Duché P. Metabolic syndrome in obese children and adolescents: dichotomous or continuous? *Metab Syndr Relat Disord.* 2009;7:549-56.

[76] Eisenmann JC. On the use of a continuous metabolic syndrome score in pediatric research. *Cardiovascular Diabetology* 2008;7:17.

[77] Kahn R, Buse J, Ferrannini E, Stern M. The metabolic syndrome: time for a critical appraisal: joint statement from the American Diabetes Association and the European Association for the Study of Diabetes. *Diabetes Care* 2005;28(9):2289-304.

[78] Osei K, Rhinesmith S, Gaillard T, Schuster D. Impaired insulin sensitivity, insulin secretion, and glucose effectiveness predict future development of impaired glucose tolerance and type 2 diabetes in pre-diabetic African Americans: implications for primary diabetes prevention. *Diabetes Care* 2004;27(6):1439-46.

[79] Morrison JA, Glueck CJ, Horn PS, Wang P. Childhood predictors of adult type 2 diabetes at 9- and 26-year follow-ups. *Arch Pediatr Adolesc Med* 2010;164(1):53-60.

[80] Martin BC, Warram JH, Krolewski AS, Bergman RN, Soeldner JS, Kahn CR. Role of glucose and insulin resistance in development of type 2 diabetes mellitus: results of a 25-year follow-up study. *Lancet* 1992;340(8825):925-9.

[81] Weyer C, Hanson RL, Tataranni PA, Bogardus C, Pratley RE. A high fasting plasma insulin concentration predicts type 2 diabetes independent of insulin resistance: evidence for a pathogenic role of relative hyperinsulinemia. *Diabetes* 2000;49(12):2094-101.

[82] Weiss R, Taksali SE, Tamborlane WV, Burgert TS, Savoye M, Caprio S. Predictors of changes in glucose tolerance status in obese youth. *Diabetes Care* 2005;28(4):902-9.

[83] Kynde I, Heitmann BL, Bygbjerg IC, Andersen LB, Helge JW. Hypoadiponectinemia in overweight children contributes to a negative metabolic risk profile 6 years later. *Metabolism* 2009;58(12):1817-24.

[84] Ridker PM, Rifai N, Rose L, Buring JE, Cook NR. Comparison of C-reactive protein and low-density lipoprotein cholesterol levels in the prediction of first cardiovascular events. *N Engl J Med* 2002;347(20):1557-65.

[85] Pradhan AD, Manson JE, Rifai N, Buring JE, Ridker PM. C-reactive protein, interleukin 6, and risk of developing type 2 diabetes mellitus. *Jama* 2001;286(3):327-34.

In: Childhood Obesity: Risk Factors, Health Effects...
Editor: Carol M. Segel

ISBN: 978-1-61761-982-3
©2011 Nova Science Publishers, Inc.

Chapter V

Migrant Status - An Important Risk Factor for Developing Obesity During Childhood

*Sylvia Kirchengast**
University of Vienna, Austria

Abstract

Obesity rates are increasing worldwide at an alarming rate. Today more than 1.1 billion adults worldwide are classified as overweight, 312 million of them as obese. Especially dramatic is the situation among children and adolescents. According to the WHO overweight or obesity affects one in ten children or adolescents worldwide. In Europe one in five children can be classified as overweight or obese. Although overweight and obesity are becoming increasingly prevalent nearly all over the Industrialised world special risk groups can be identified. In the United States as well as in Europe migrant status seems to increase the risk for developing overweight or obesity during childhood an adolescence dramatically. Migration per se does not lead to increased obesity rates but migrant status is still associated with a low socioeconomic position, adverse nutritional habits and a low level of physical activity. Additionally religious and social restrictions increase the problem. In this paper the impact of migration on obesity prevalence among Turkish children living in Vienna Austria, is focused on. Beside methodological problems the impact of migration on behavioural and social factors during childhood is discussed. Within the European Union the majority of migrants suffering from excessive body weight originate from Mediterranean countries, but also from the Near and the Middle East. Among this subgroup an effective prevention of childhood obesity seems to be extraordinary difficult because cultural and religious norms make any prevention concept containing changes in eating behaviour and physical activity patterns difficult.

* Univ.Prof.dr.Sylvia Kirchengast, University of Vienna, Department for Anthropology, Althanstrasse 14, A-1090 Vienna, Austria, Email: sylvia.kirchengast@univie.ac.at

Introduction

At the beginning of the third millennium a rising prevalence of overweight and obese children and adolescents is seen in developed as well as developing and threshold countries (Kimm & Obarzanek 2002, Raman 2002, Dehghan et al. 2005, Kosti & Panagiotakos 2006, Hossein et al. 2007). According to the WHO overweight or obesity affects one in ten children or adolescents worldwide (WHO 2006). In Europe one in five children can be classified as overweight or obese (Lasserre et al. 2007). This tendency is really a dramatic one because childhood obesity is not only an aesthetic problem which may result in social stigmatization of affected children (Kraig & Keel 2001), childhood obesity is a multisystem disease with potentially devastating consequences (Ebbeling et al. 2002). As with obesity in adults, childhood obesity is acknowledged to be one of the most important risk factors for hypertension, diabetes mellitus type II, or abnormal lipid profiles during childhood as well as during later life (Geiss et al. 2001, McGee 2005, Lob-Corzilius 2007). Mossberg (1989) reported a significantly higher mortality among overweight or obese children in Sweden caused predominantly by cardiovascular diseases. Beside this well documented somatic risk factors childhood obesity also increase the risk for psychic and emotional morbidity during childhood, adolescence and during later life (Latner & Stunkard 2003). Furthermore obesity in childhood is a significant predictor of obesity in adulthood (Caprio et al. 2008). Prevention and effective treatment of childhood obesity are therefore very important public health issues today, however before an adequate prevention program or effective treatment can be introduced a profound analysis of causes and risk factors of childhood obesity is absolutely necessary.

Causes and Risk Factors of Childhood Obesity

Obesity is a multifactorial caused condition. Various intrinsic and extrinsic factors may contribute to the risk to become obese. Among intrinsic factors a genetic predisposition and the hormonal situation are can have a great effect on the individual risk to develop obesity during childhood (Ebbeling et al. 2002, Raman 2002). Since the last decade of the 20th century several genetic mutations that can cause obesity during childhood have been identified (Farooqi & O'Rahilly 2000, Farooqi 2005, Weiss & Kaufman 2007), furthermore the results of twin and adoption studies plead for a strong effect of genetic factors on the variation of the body mass index (Silventoinen et al. 2010). From a physiological point of view hormones influence the development of obesity during childhood, vice versa adipose tissue is a source of hormones. Hypothalamic or pituitary dysfunction results in altered hormone secretion. A decreased secretion of growth hormone, sex steroids or thyroid hormones is also associated with an increased risk of obesity (Weiss et al. 2004). Another intrinsic factor contributing to childhood obesity is prenatal overnutrition. It is assumed that that maternal overweight increases the transplacental nutrients transfer, a condition which may result in permanent changes in appetite, neuroendocrine functions and metabolism (Ebbeling et al. 2002). Another early risk factor for developing obesity during childhood is bottle-feeding. It could be shown that bottle-fed children had an increased risk to become obese in comparison to their breast fed counterparts (Toschke et al. 2007.). Beside these intrinsic factors the effects of extrinsic factors on the increasing prevalence of childhood obesity are discussed. Socioeconomic factors have a dramatic influence on the risk to become obese or not. Disadvantaged groups characterized by a low socioeconomic status, poverty and a low educational level are at greatest risks for childhood obesity in Industrialised countries (Schneider 2000, Haas et al. 2003, McLaren 2007). In these countries a so called obesogenic environment prevail (Ulijaszek 2007). An obesogenic environment is generally characterized by food security i.e. the physical, social and economic access to sufficient, safe and nutritious food, but also by reduced physical activity patterns and a sedentary lifestyle. Especially fatal consequences of an obesogenic environment are observable among populations in transition. Acculturation and processes of modernization often result in the development of obesogenic environments. The result is the dramatically increase of obesity among populations in transition (Perez-Escamilla & Putnik 2007). This is mainly due to a change in eating habits, nutritional preferences. This means an increase in caloric intake and a decrease in physical activity (Galal 2002). A comparable effect can be observed among children and adults with a background of migration. The analysis of the impact of migration or background of migration on childhood obesity is the topic of the present paper.

Factor Minority Status and Migration

As pointed out above, socioeconomic factors are strongly related with the prevalence of childhood obesity. In general a low socioeconomic status of the family, low income and low educational levels as well as adverse living patterns increase the risk for getting obese during childhood. This is especially true of Industrialised countries, while in developing countries a higher likelihood of obesity among children in higher socioeconomic strata were found (McLaren 2007). Within developed countries minority status is an additional risk factor (Kumanyika 1994, 2008, Gordon-Larsen et al. 2003). In the United States ethnic minorities are predominantly African Americans, American Indians, Pima, Polynesians and Hispanic people, ethnicities who lived since a long time in the United States. This is also true of European countries with a long history of colonialism such as the United Kingdom, where ethnic minorities mainly originate from India, Pakistan or African countries. In many other European countries in contrast members of minority groups have predominantly a background of migration since the 1960ties. In these countries migration seems to be a major risk factor for childhood obesity. Although often suggested, migration is nothing new in the history of Homo sapiens. In contrast, migration of individuals and whole populations is an ancient phenomenon among Homo sapiens and his ancestors (Mascie-Taylor & Lasker 1988, Fix 1999). Today several push factors such as a lack of resources, poverty or war but also pull factors such as better educational or job opportunities but also marriage promote residential mobility and transnational migration worldwide. Migration and increased mobility however are not only of growing social, political and demographic importance, the process of migration is also an important factor of health and disease (Rogerson & Han 2002, Vissandjee et al. 2004, Uitewaal et al. 2004 a,b, Gushalk & MacPherson 2006, Misra & Ganda 2007). Once migrants have moved to their host country, they are often a minority with a lower social status than the host population (Kilaf 2004). Beside the low socioeconomic position and the minority status disparities in nutritional habits, physical activity patterns and health related behaviour between migrants and host population, seem to be associated with increased chronic distress, a higher morbidity and an increased mortality among people with a background of migration (Dotevall et al. 2000, Landman & Cruishank 2001, Bongard et al. 2002, Haas et al. 2003, Koehn 2006, Ujicic-Voortman et al. 2009). A special health problem of migrants is the high prevalence rate of overweight and obesity, especially among women and children or adolescents (Brussard et al.2001, Hoppichler & Lechleitner 2001, Kilaf 2004, Kirchengast & Schober 2005, 2006). As mentioned above overweight and obesity are becoming increasingly prevalent nearly all over the Industrialised world at the beginning of the third millenium (Martinez 2000, Livingstone 2001, Tomkins 2001), nevertheless special risk groups were described. Among these risk groups, people with a background of migration are frequently found. In the United States as well as in Europe to be migrant and belong to a ethnic minority group represents a special risk factor for developing overweight or obesity during childhood and adolescence (Popkin & Urdy 1998, Brussard et al. 2001, Fredriks et al. 2003, Gordon-Larsen et al. 2003a,b, Erb & Winkler 2003, Kirchengast & Schober 2005, 2006, Caprio et al. 2008, Fitzgibbon & Beech 2009) but also during adulthood (Dijkshoorn et al. 2008, Wolin et al. 2009). In the United States especially high rates of childhood obesity among Hispanic, African American and Native American children are reported (Kumanyika 1994, Gordon-Larsen et al. 2003, Perez-Escamilla & Putnik 2007, Caprio et al. 2008) and the risk to remain obese as adults is especially high among minority children (Caprio et al. 2008). In Central and Northern Europe in contrast, a high prevalence of overweight or obesity was mainly found among migrants originating from Mediterranean countries, such as Greece, Italy, Turkey or Morocco (Dotevall et al. 2000, Frederiks et al. 2003, Erb & Winkler 2003, Kirchengast & Schober 2005, 2006, Dijkshoorn et al. 2008).

In the present study the very specific situation of Vienna, Austria is considered. There is no doubt that Austria is today a country of immigration, although Austria can not be considered as a typical country of immigration, like Canada, United States or Australia. Although Austria had been one of the largest Empires in Europe up to the end of the First World War, Austria never had any colonies. Nevertheless various ethnic minorities lived in Vienna at the beginning of the 29th century. These minority people originated from the eastern and southern parts of the Austrian-Hungarian Monarchy. So Polish people, Czechs, Croatians, Slovenians, Hungarians, Macedonians, Serbs and Italians lived in Vienna. After the First World War only the very small German speaking western Part of Austria-Hungary remained as Austria and the populations became relatively homogeneous. Since the 1960s, however an increasing number of labour migrants from Former Yugoslavia and Turkey migrated to Austria (Kilaf 2004). At the end of the eighties and during the early nineties of the last century migration resulted primarily from family reunion on the one hand and from political refugees especially from former Yugoslavia, on the other hand. Today 17% of the Austrian inhabitants have a background of migration the majority of them originate from Former Yugoslavia and Turkey. These people with a background of migration are the new ethnic minority groups of Austria. The aim of the present study was to analyse prevalence of overweight and obesity among children and adolescents originating from Turkey.

Viennese Study: Obesity among Turkish Migrant Children in Vienna

Data Collection and Subjects

All together 2940 children ageing, 6, 10 and 15 years were enrolled in the present study. Data collection took place in strong co-operation with the Viennese school medical authority. Forty six public schools (so-called Hauptschulen) of Vienna (two from each of the 23 districts of Vienna) were randomly selected from the Viennese school authority to participate in the project. For the present study three age groups of children originating from Turkey were analysed. Age group 1 comprised 272 six year old children. This first examination took place immediately before school entry. Age group 2 comprised 275 ten year old children. This examination took place at the end of primary school. Age group 3 comprised 415 boys and girls. This last examination took place at the age of fifteen years, short before the school attendance ends. The data files including all information concerning stature and body weight are stored by the Viennese school medical officers. We were allowed to use the data of the Viennese school medical office. It is very difficult to define migrant status because different definitions to identify a migrant are used (Brussard et al. 2001). In the present study it was decided to use the definition recommended by United Nations Economic Commission for Europe (UNECE 2006). According to this definition all persons whose both parents were born outside the country had still migrant status. These migrants can be classified as first generation migrants if they were born also Austria or as second generation migrants if they were born in Austria. In the present sample migrant status was obtain by the country of birth of both parents. Therefore only children whose both parents were born in Turkey were included in the present sample.

As controls 343 boys and 268 girls aging six years, 340 boys and 260 girls aging ten years and 428 boys and 339 girls aging fifteen years were enrolled in the present study. All subjects of the control group originated from Austria. Both groups of children and adolescents, Turkish as well as Austrian ones belonged to the lower social stratum of Vienna. This low social status was determined by attendance to a public so called Hauptschule as well as by socioeconomic parameters such as educational level and profession of the parents.

Somatometric Parameters

In Austria school children are medically examined every year. Beside medical data, data regarding stature height and body weight of all school children are collected obligatory by special educated personal of the medical school authority. Stature is measured with an anthropometer to the nearest millimetre. Weight was recorded with a scale precise to +/- 100g. The children and adolescents were measured without shoes wearing underwear only.

Weight Status

Weight status was determined by using the body mass index (BMI) kg/m^2. In order to classify weight status categories different classification schemes for children and adults were used.

Weight Status Classification among Children and Adolescents

Although the BMI is increasingly used for the diagnosis of obesity, overweight and underweight in childhood and adolescence (Zarfl & Elmadfa 1995), up to now there is a lack of European standards (Cole et al. 1995, 2000, Chinn & Rona 2002). In a country of migration, like Austria, the dilemma is that no growth and body mass references for immigrant children in Austria exists up to now (Kirchengast & Schober 2009). This lack of growth and body mass data is not only an Austrian problem it is still a problem in several typical countries of immigration, such as the Netherlands (Fredriks et al. 2003). In the present study in a first step this problem was focused on (Part one of the result section).

For the analyses of prevalence of obesity among Austrian children with a background of migration from Turkey the percentiles of the body mass index published by Kromeyer - Hauschild et al. (2001) for central Europe were used. According to the recommendations of Kromeyer-Hausschild et al (2001) and the European Childhood obesity group (Zwiauer & Wabitsch 1997) the 90 percentile as cut-off for overweight and the 97 percentile as cut-off for obesity were defined.

Results of the Viennese Study

1. Problem of Obesity Classification of Children with a Background of Migration

According to the WHO obesity is defined as, a condition with excessive fat accumulation (WHO 1998). Techniques to estimate body fat in a reliable manner, however, are expensive and not practicable in large epidemiological surveys. Therefore the body mass index (BMI) the simple weight to height^2 (kg/m^2) ratio is used widely. For optimal monitoring of a population up-to-date reference BMI data on representative samples from the population are necessary. During the last two decades numerous reference curves for different countries or populations are published (Cole et al. 1990, Shang et al. 2005, de Onis et al. 2007). These national BMI reference curves for children and adolescents however are not very practicable for so called multiethnic societies resulting from globalization and transnational migration (Fredriks et al. 2003). In Austria as in many other European countries the ongoing phenomenon of immigration makes the situation for paediatricians and anthropologists complicated. They have to decide what kind of references should be used: BMI reference curves for a representative sample of the whole (multiethnic) population or should separate reference curves for each larger subpopulation be created? Another option is the use reference curves established for the country or region of origin of each subpopulation. In order to overcome this dilemma, international standard definitions of overweight and obesity such as those of the WHO (WHO 2006) or of Cole et al (2000) were established. Because of a lack of recent BMI curves of the Austrian population, Austrian paediatricians and anthropologists use the percentile curves published by Kromeyer-Hauschild et al (2001). But are these curves appropriate to describe the weight status of children originating from Turkey? Cole et al. (2000) published internationally based cut-offs. However this international standard was also critized and some limitation were discussed (Wang & Wang 2000, Chinn & Rona 2002). But problems do not only arise from international comparisons. A special problem was pointed out by Frederiks et al. (2003): the dilemma of countries with a high proportion of children with a background of migration. In the case of Fredriks et al. (2003) the situation of Turkish children in the Netherlands was described. In such multicultural and multiethnic societies it is not clear which kind of reference charts should be used to evaluate weight status among children and adolescents: reference charts from a representative sample of the whole multiethnic society or separate charts for each subpopulation. Frederiks et al. (2003) suggested the development of separate charts for each subpopulation. In contrast the study of Stellinga–Boelen et al. (2007) used Dutch reference charts for children of Asylum seekers from different origin and reported no problems by using Dutch national reference curves. Another possibility to overcome this problem is the use of international standards, however the validity of international standard references was also discussed critically (Bozkirli et al. 2007, Huetra et al. 2007). In case of the Austrian situation in a previous study (Kirchengast & Schober 2009) three different reference charts were compared based on the BMI data of Turkish children in Austria. The in Austria used percentile curves of Kromeyer-Hauschild et al. (2001) were compared with the international standards published by Cole et al. (2000) and the reference curves for Turkish children in Anatolia published by Özer (2002). These Özer charts had higher cut offs for overweight and obesity than all other reference charts. Therefore the Özer charts yielded the lowest prevalence of obesity and overweight for girls as well as boys of the present sample. The highest prevalence of overweight and obesity was using the charts of Kromeyer-Hauschild et al. (2001) and Cole et al. (2000). The national reference curve of Kromeyer-Hauschild et al. (2001) for German children and adolescents showed also the correspondence with the international standards published by Cole et al. (2000). But what does this mean for practice? Using international standards and National standards for German children for Austrian children with a background of migration from Turkey differ significantly from those published for Turkish children in Turkey. The problems of anthropologists as well as paediatricians monitoring weight status during childhood and adolescence still remain. Since it was not possible to develop new reference curves for Austrian children with a background of migration from Turkey in the present study the percentile curves published by Kromeyer-Hauschild et al. (2001) were used.

2. Prevalence of Overweight and Obesity among Children and Adolescents

The prevalence of overweight and obesity was generally high among Turkish children and adolescents as well as among the Austrian controls. As to be seen in table 1, Turkish girls surpassed their Austrian counterparts in the prevalence of overweight and obesity. This was true of all age groups. Statistically significant differences (p <0.01) in the prevalence of overweight and obesity however were found for the oldest age group (15 years) only.

Table 1. Weight status (BMI percentiles) according to age, sex and ethnicity among children and adolescents

		girls		Sig	boys		Sig
		Turkish	Austrian		Turkish	Austrian	
6 years							
	Overweight	12.5%	9.3%	n.s.	9.7%	7.0%	n.s.
	obese	12.4%	10.8%		9.0%	10.2%	
10 years							
	Overweight	16.5%	15.4%	n.s.	10.3%	11.4%	n.s.
	obese	13.4%	10.4%		8.9%	11.2%	
15 years							
	Overweight	16.4%	15.3%	p<0.01	14.3%	12.1%	n.s.
	obese	12.5%	8.6%		9.4%	10.7%	

The relative risk of being overweight or obese was always higher among Turkish girls (table 2). Among Turkish boys in contrast, the prevalence of obesity was always lower than among their Austrian counterparts. Regarding the prevalence of overweight 6 as well as 15 year old Turkish boys surpassed the Austrian controls. Therefore the relative risk of being overweight was only higher among 6 and 15 year old Turkish boys in comparison with their Austrian controls (table 2).

Table 2. Relative risk of being overweight or obese in comparison with Austrian children of comparable age

	Turkish girls	Turkish boys
Age	Odds ratio (CV 95%)	Odds ratio (CV 95%)
overweight		
6 years	1.11 (0.982-1.440)	1.12 (0.873-1.441)
10 years	1.04 (0.850-1.264)	0.84 (0.737-0.967)
15 years	1.26 (1.003- 1.591)	1.06 (0.889-1.273)
obese		
6 years	1.17 (0.687-1.977)	0.97 (0.808-1.167)
10 years	1.11 (0.864-1.419)	0.90 (0.758-1.077)
15 years	1.41 (0.877-2.258)	0.59 (0.806-1.141)

Discussion

In the present paper the specific situation of the second largest immigrant group, people originating from Turkey, in Austria was focused on. Over the last few decades Austria has undergone a change, from a relatively homogeneous society to a so called multicultural one. In the year 2006, according to National census, 134 299 Turkish migrants, who did not have Austrian citizenship lived in Austria (Statistik Austria 2006). About 100000 Turkish people had the Austrian citizenship at this time. People with a background of migration from Turkey are therefore an important minority group in Austria. The majority of these Turkish migrants live in Vienna, where more than 20% of the inhabitants have a background of migration today. So the Austrian society is increasingly characterized by many different languages, religions and cultural traditions. This trend has brought the health and health problems of different ethnic groups into focus. As described above one important health risk factor which is frequently found among migrants is obesity (Brussard et al. 2001, Hoppichler & Lechleitner 2001, Fredriks et al. 2003, Erb & Winkler 2003). In European countries especially migrants originating from Mediterranean region and from the near and middle east show an extremely high prevalence of overweight and obesity (Fredriks et al. 2003). This is especially true of children and adolescents as well as women (Fredriks et al. 2003, Kirchengast & Schober 2005, 2006, Dijkshoorn et al. 2008). The results of the present study corroborate these reports. While Turkish boys were not more frequently classified as overweight or obese than their Austrian counterparts, the prevalence of overweight and obesity was dramatically higher among Turkish girls in comparison with their Austrian counterparts.

In general obesity is not uncommon among Turkish people in Turkey as well as in European countries (Erem et al. 2001, 2004, Dinc et al. 2006, Dijkshoorn et al. 2008, Oguz et al. 2008, Papandreou et al. 2008, Gültekin et al. 2009). In Turkey according to Iseri and Arslan (2008) 56% of the adult population is overweight or obese. Oguz et al. (2008) reported a prevalence of overweight of 36.0% and a prevalence of obesity of 30.4% for the adult Turkish population. 41.4 % of the adult population of Trabzon was classified as overweight and 19.2% as obese (Erem et al. 2004). An extraordinary high prevalence of overweight and obesity is also found among Turkish migrants in European countries: Dijkshorn et al. (2008) reported a prevalence of overweight/obesity of more than 80% among Turkish migrants in the Netherlands. Nearly all authors reported higher obesity rates for among women than for men. These results are comparable to those of the present study, although all studies mentioned above focused on adults only. However we have to be aware that all studies mentioned above focused on adults only.

A little bit different is the situation among children and adolescents, while some authors report low rates (0.9% to 3.8%) (Aygun et al. 1997, Özer 2007) or 3.7% to 12.2% (Discigil et al. 2009) of overweight and obesity among children and adolescents in Turkey, extremely high rates re reported for Turkish migrant children in European countries, such as 23.4 - 30.2% for Turkish children in the Netherlands (Fredriks et al. 2005). These percentages are comparable to those of the present sample. But what are the reasons for this extraordinary high prevalence of overweight and obesity among Turkish people especially among girls in Austria?

Beside genetic factors (Kimm 2003) obesity seems to be largely caused by an environment that promotes excessive food intake and discourages physical activity (Hill & Peters 1998). Focusing on the situation of Turkish migrants in Vienna, the impact of sociocultural parameters should not be underestimated. From a social point of view minority group status, low socioeconomic status or to be a migrant are mentioned to be important risk factors to become overweight or obese (Haas et al. 2003). A low socioeconomic status and/or discrimination by ethnicity may result in increased stress, which has a direct effect on the hypothalamic-pituitary-adenal axis. The result is an elevated cortisol level, which may promote the development of obesity (Caprio et al. 2008). The fact, that in the present study

Austrian children, predominantly girls, are less frequently overweight or obese is in accordance with the results from other European countries: Turkish migrant children are more frequently overweight or obese than children of the host population (Fredriks et al. 2005, Will et al. 2005). Differences in socioeconomic status cannot be assumed to be responsible for these differences between the ethnic groups, because in the present study all children belonged to the lower socioeconomic stratum of Vienna. A low social or socioeconomic status was recently declared to be one of the major risk factors for developing overweight during childhood (Jain et al.2001, Haas et al. 2003, Lioret et al. 2007), although low income children have historically been regarded as at-risk for undernutrition. In Turkey this true even today, but in most Industrialised countries the opposite is true. On the one hand, causes for this trend are economic ones, food characterised by high energy density, such as sweets or fats, is much cheaper than vegetables or fresh fruits (Caprio et al. 2008). Therefore in the modern world poverty is associated with overweight and obesity (Drewnowski & Specter 2004, Shahar et al. 2005, Lawrence et al. 2007, Pollestad et al. 2008). On the other hand physical activity is reduced among children of low socioeconomic status, especially among children with a background of migration (Green et al. 2003, Haas et al. 2003, Hosper et al. 2007). Child activity patterns as well as nutritional habits are highly influenced by culture and religious components (Shatenstein & Ghadirian 1998, Dowler 2001, Gordon-Larsen et al. 2003a,b, Green et al. 2003, Hosper et al. 2007). Culture influences preferences for and opportunities to engage physical activity, but also child-feeding patterns und nutritional preferences (Carpio et al. 2008, Kral & Rauh 2010). High energy diet on the one hand and reduced physical activity characterised by long time watching TV (Gordon-Larsen et al. 2002) on the other hand seemed to be the main reason for the high levels of overweight and obesity among migrant children, especially originating from Turkey, observed in several European countries (Brussard et a. 2001, Erb & Winkler 2003, Fredriks et al. 2003). Furthermore culture can influence the perception of risk associated with obesity (Fitzgibbon & Beech 2009). Low income mothers, but also mothers with a background of migration are not worried about the overweight of their offspring (Jain et al. 2001). According to these women, their offspring is healthy and well-nourished, not overweight or obese. This may be due to the fact that up to now overweight or obesity during childhood in Turkey is phenomenon predominantly found in social middle and high income class (Manios et al. 2004). Therefore overweight is culturally positively interpreted and not seen as an important long time health risk by the majority of the Turkish population. The extremely high prevalence of overweight and obesity among adult women may be explained in a similar way. The majority of Turkish women and girls in Vienna underlie strict cultural and religious pressures, characterised by extremely low physical activity outside the household during leisure time (Gordon-Larsen et al. 2002, Kilaf 2004, Hosper et al. 2007). Furthermore typical nutritional habits promote weight gain and the development of obesity (Darmon & Khlat 2001, Dowler 2001, Mellin-Olsen & Wandel 2005, Lawrence et al. 2007). Although overweight and obesity reduce health related quality of life dramatically, even among obese Turkish children and women (Dinc et al. 2006) women are not worried about being obese or about obesity among their children. This is especially true of women with a low educational level and low degree of acculturation or integration in the Austrian society. This association between low educational level and a high prevalence of obesity was also described for Turkish women in Istanbul (Tanyolac et al. 2008).

We can conclude that obesity is not only a medical problem it is also a socioeconomic and cultural one (Ulijaszek & Lofink 2006). In the future it is absolutely necessary that health professionals in all multicultural countries consider these aspects in their concepts.

References

Aygun D, Akarsu S, Yenioglu H, Kocabay K, Guvenc H 1997. Prevalence of obesity in adolescents in Elazig city. *Firat Tip Dergisi* 1: 167-170.

Bongard S, Pogge SF, Arslaner H, Rohrmann S, Hodapp V 2002. Acculturation and cardiovascular reactivity of second generation Turkish migrants in Germany. *J Psychosomatic Res* 53: 795-803.

Bozkirli E, Ertorer ME, Bakiner O, Tutunc NB, Demirag NG 2007. The validity of the world health organisation´s obesity body mass index criteria in a Turkish population: a hospital-based study. *Asia Pac J Clin Nutr* 16: 443-447.

Brussard JH, Erp-Baart MA van, Brants HAM, Hulshof KFAM, Löwik MRH 2001. Nutrition and health among migrants in the Netherlands. *Public Health Nutrition 4*: 659-664.

Caprio S, Daniels SR, Drewnowski A, Kaufman FR, Palinkas LA, Rosenbloom AL, Schwimmer JB 2008. Influence of race, ethnicity, and culture on childhood obesity: implications for prevention and treatment. *Diabetes Care* 31: 2211-2221.

Chinn S, Rona RJ 2002. International definitions of overweight and obesity for children, a lasting solution? *Ann Hum Biol* 29: 306-313.

Cole TJ, Freeman JV, Preece MA 1995. Body mass reference curves for the UK, 1990. *Arch Dis Child* 73: 25-29.

Cole TJ, Bellizzi MC, Flegal KM, Dietz WH. Establishing a standard definition for child overweight and obesity worldwide: international survey. *Brit Med J* 2000; 320: 1-6.

Darmon N, Khlat M 2001. An overview of the health status of migrants in France, in relation to their dietary practices. *Public Health Nutrition* 4: 163-172.

Dehghan M, Akhtar-Danesh N, Merchant AT. Childhood obesity, prevalence and prevention. *Nutrion J* 2005; 4: 24-32

Dijkshoorn H, Nierkens V, Nicolaou M 2008. Risk groups for overweight and obesity among Turkish and Moroccan migrants in The Netherlands. *Public Health* 122: 625- 630

Dinc G, Eser E, Saatli GL, Cihan UA, Oral A, Baydur H, Ozcan C, 2006. The relationship between obesity and health related quality of life of women in a Turkish city with a high prevalence of obesity. *Asia Pacific J Clin Nutr* 15: 508-515.

Discigil G, Tekin N, Soylemez A 2009. Obesity in Turkish children and adolescents: prevalence and non-nutritional correlates in an urban sample. *Child: Care, Health and Development* 35: 153-158.

Dotevall A, Rosengren A, Lappas G, Wilhelmsen L 2000. Does immigration contribute to decreasing CHD incidence? Coronary risk factors among immigrants in Göteborg, Sweden. *J Internal Med* 247: 331-339.

Dowler E 2001. Inequalities in diet and physical activity in Europe. *Public Health Nutrition* 4: 701-709.

Drewnowski A, Specter SE 2004. Poverty and obesity, the role of energy density and energy costs. *Am J Clin Nutr* 79: 6-16.

Ebbeling CB, Pawlak DB, Ludwig DS 2002. Childhood obesity: public-health crisis, common sense. *Lancet* 360: 473.

Erb J, Winkler G 2003. Rolle der Nationalität bei Übergewicht und Adipositas bei Vorschulkindern. *Monatsschrift Kinderheilkunde* 152: 291-298.

Erem C, Yildiz R, Kavgaci H, Karahan C, Deger O, Can G, Telatar M 2001. Prevalence of diabetes, obesity and hypertension in a Turkish population (Trabzon city). *Diabetes Res Clin Pract* 54: 203-208

Erem C, Arslan C, Hacihasanoglu A, Deger O, Topbas M, Ukinc K, Ersöz HÖ, Telatar M 2004. Prevalence of obesity and associated risk factors in a Turkish population (Trabzon City, Turkey). *Obes Res* 12: 1117-1127.

Farooqi IS, O'Rahilly S 2000. Recent advances in the genetics of severe childhood obesity. *Arch Dis Child* 83: 31-34.

Farooqi IS 2005. Gentic and hereditary aspects of childhood obesity. *Best Practice Clin Endocrinol Metab* 19: 359-374.

Fitzgibbon ML, Beech BM 2009. The role of culture in the context of school based BMI screening. *Pediatrics* 124: S50-"62.

Fix AG 1999. Migration *and colonization in human microevolution.* Cambridge University Press, Cambridge

Fredriks AM, van Buuren S, Jeurissen SER, Dekker FW, Verloove-Vanhorick SP, Wit JM 2003. Height, weight, body mass index and pubertal development reference values for children of Turkish origin in the Netherlands. *Eur J Pediatrics* 162: 788-793.

Fredriks AM, van Buuren S, Hira Sing RA, Wit JM, Verloove-Vanhorick SP 2005. Alarming prevalence of overweight and obesity for children of Turkish, Moroccan and Dutch origin in The Netherlands according to international standards. *Acta Paediatrica* 94: 496-498.

Galal OM 2002. The nutrition transition in Egypt: obesity, undernutrition and food consumption context. *Public Health Nutrition* 5 141-148.

Geiss HC, Parhofer KG, Schwandt P 2001. Parameters of childhood obesity and their relationship to cardiovascular risk factors in healthy prepubescent children. *Int J Obes* 25: 830-837.

Gordon-Larsen P, Adair LS, Popkin BM 2003a. The relationship of ethnicity, socioeconomic factors and overweight in US adolescents. *Obes Res* 11: 121-129.

Gordon-Larsen P, Mullan K, Ward DS, Popkin BM 2003b. Acculturation and overweight-related behaviours among immigrants to the US, the national longitudinal study of adolescent health. *Soc Sci Med* 57: 2023-2034.

Gordon-Larsen P, Adair LS, Popkin BM 2002. Ethnic differences in physical activity and inactivity patterns and overweight status. *Obes Res* 10: 141-149.

Green J, Waters E, Haikerwal A, O'Neill CO, Raman S, Booth ML, Gibbons K 2003. Social, cultural and environmental influences on child activity and eating in Australian migrant communities. *Child Care Health Develop* 29: 441-448.

Gültekin T, Ozer BK, Akin G, Bektas Y, Sagir M, Gülec E 2009. Prevalence of overweight and obesity in Turkish adults. *Anthrop Anz* 6: 205-212

Gushalk BD, MacPherson DW 2006. The basic principles of migration health: Population mobility and gaps in disease prevalence. *Emerging Themes Epidemiol* 3:1-10.

Haas JS, Lee LB, Kaplan CP, Sonneborn D, Phillips KA, Liang SY 2003. The association of race, socioeconomic status and health insurance status with the prevalence of overweight among children and adolescents. *Am J Public Health* 93: 2105-2110.

Hill JO, Peters JC 1998. Environmental contributions to obesity epidemic. *Science* 280: 1371-1374.

Hoppichler F, Lechleitner M 2001. Counseling programs and the outcome of gestational diabetes in Austrian and Mediterranean Turkish women. *Patient Education Counseling* 45: 271-274.

Hosper K, Klazinga NS, Stronks K 2007. Acculturation does not necessarily lead to increased physical activity during leisure time: a cross-sectional study among Turkish young people in the Netherlands. *Public Health* 7: 230-239.

Hossain P, Kawar B, El Nahas M 2007. Obesity and diabetes in the developing world – a growing challenge. *New England J Med* 356: 213-215.

Huerta M, Gdalevich M, Tlashadze A, Scharf S, Schlezinger M, Efrati O, Bibi H. 2007. Appropriateness of US and international BMI-for age reference curves in defining adiposity among Israeli school children. *Eur J Pediatr* 166: 573-578.

Iseri A, Arslan N 2008. Obesity in adults in Turkey: age and regional effects. *Eur J Public Health* 19: 91-94.

Jain A, Sherman SN, Chamberlin LA, Carter Y, Powers SW, Whitaker RC 2001. Why don't low income mothers worry about their preschoolers being overweight? *Paediatrics* 107: 1138-1146.

Kilaf E 2004 *Turkish migration to Austria and women's health.* PHD Thesis University of Vienna

Kimm SYS 2003. Nature versus nurture I childhood obesity, a familiar old conundrum. *Am J Clin Nutr* 78:1051-1052

Kimm SYS, Obarzanek E 2002. Childhood obesity: A new pandemic of the new millennium. *Paediatrics* 110: 1003-1007

Kirchengast S, Schober E 2005. To be an immigrant: a risk factor for developing overweight and obesity during childhood and adolescence. *J biosoc sci* 38: 695-705

Kirchengast S, Schober E 2006. Obesity among female adolescents in Vienna, Austria – the impact of childhood weight status and ethnicity. *Brit J Obstet Gynecol*113:1188-1194.

Kirchengast S, Schober E. 2009. Growth charts in a globalizing world: A new challenge for anthropologists and paediatricians? *Acta Medica Lituanica* 16: 76-82

Koehn PH 2006. Globalization, migration health, and educational preparation for transnational medical encounters. *Globalisation and Health* 2: 2-18.

Kosti RI, Panagiotakos DB 2006. The epidemic of obesity in children and adolescents in the world. Cent Eur J Public Health 14: 151-159

Kraig KA, Keel PK 2001. Weight based stigmatisation in children. Int J Obes 25: 1661-1666.

Kral TVE, Rauh EM 2010. Eating behaviour of children in the context of their family environment. *Physiol Behav* (in press)

Kromeyer-Hausschild K, Wabitsch M, Kunze D, Geller F, Geiß HC, Hesse V, von Hippel A, Jäger U, Korte W, Menner K, Müller G, Müller JM, Niemann-Pilatus A, Remer T, Schäfer F, Wittchen HU, Zabransky S, Zellner K, Ziegler A, Hebebrand J 2001. Perzentile für den Body –mass Index für das Kindes- und Jugendalter unter Heranziehung verschiedener deutscher Stichproben. *Monatsschrift Kinderheilkunde* 149: 807-818.

Kumanyika SK 1994. Obesity in minority populations: an epidemiologic assessment. Obes Res 2: 166-182.

Kumanyika SK 2008. Environmental influences on childhood obesity: Ethnic and cultural influences in context. *Physiol & B*ehav 94: 61-70.

Landman J, Cruickshank JK 2001. A review of ethnicity, health and nutrition related diseases in relation to migration in the United Kingdom. *Public Health Nutr* 4: 647-657.

Lasserre AM, Chiolero A, Paccaud F, Bovert P 2007. Worldwide trends in childhood obesity. *Swiss Med Weekly* 137: 157-158

Latner JD, Stunkard AJ 2003. Getting worse: the stigmatisation of obese children. *Obes Res* 11: 452-456.

Lawrence JM, Devlin E, Macaskill S, Kelly M, Chinouya M, Raats MM, Barton KL, Wrieden WL, Shepherd R 2007. Factors that affect the food choices made by girls and young women, from minority ethnic groups, living in the UK. *J Hum Nutr Diet* 20: 311-319.

Lioret S, Maire B, Volatier JL, Charles MA 2007. Child overweight in France and its relationship with physical activity, sedentary behaviour and socioeconomic status. *Eur J Clin Nutr* 61: 509-516.

Livingstone MBE 2004. Childhood obesity in Europe, a growing concern. *Public Health Nutr* 4:109-116.

Lob-Corzilius T 2007. Overweight and obesity in childhood – a special challenge for public health. *Int J Hyg Environ Health* 210: 585-589.

Manios Y, Dimitriou M, Moschonis G, Kocaoglu B, Sur H, Keskin Y, Hayran O 2004. Cardiovascular disease risk factors among children of different socioeconomic status in Istanbul, Turkey. Directions for public health and nutrition policy. *Lipids in Health Disease* 3: 11-17.

Martinez JA 2000. Obesity in young Europeans, genetic and environmental influences. *Eur J Clin Nutr* 54: S56-S60

Mascie-Taylor CGN, Lasker GW 1988. *Biological aspects of human migration* Cambridge University Press, Cambridge.

McGee DL 2005. Body mass index and mortality, a meta-analysis based on person level data from twenty-six observational studies. *Ann Epidemiol* 15: 87-97.

McLaren L 2007. Socioeconomic status and obesity. *Epidemiol Rev* 29: 29-48.

Mellin-Olsen T, Wandel M 2005. Changes in food habits among Pakistani immigrant women in Oslo, Norway. *Ethnicity Health* 4: 311-339.

Misra A, Ganda OP 2007. Migration and its impact on adiposity and type 2 diabetes. *Nutrition* 23: 696-708.

Mossberg HO 1989. 40 years follow up of overweight children. *Lancet* 2: 491-493.

Oguz A, Temizhan A, Abaci A, Kozan O, Erol C, Ongen Z, Celik S 2008. Obesity and abdominal obesity; an alarming challenge for cardiovascular risk in Turkish adults. *Anadolu Kardiyolu Derg* 8: 401-406

Onis de M, Onyango AW, Borghi E, Siyam A, Nishida C, Siekmann J 2007. Development of a WHO growth reference for school-aged children and adolescents. Bull World Health Org 85: 660-667

Özer BK 2007. Growth reference centiles and secular changes in Turkish children and adolescents. Econom Hum Biol 5: 280-301.

Papandreou C, Abu Mourad T, Jildeh C, Abdeen Z, Philalithis A, Tzanakis N 2008. Obesity in Mediterranean region (1997-2007) a systematic review. *Obes Rev* 9: 389-399

Perez-Escamilla R, Putnik P 2007. The role of acculturation in nutrition, lifestyle and incidence of type 2 diabetes among Latinos. *J Nutr* 137: 860-870.

Pollestad Kolsgaard ML, Andersen LF, Tonstad S, Brunborg C, Wangensteen T, Joner G 2008.. Ethnic differences in metabolic syndrome among overweight and obese children and adolescents: the Oslo adiposity intervention study. *Acta Paediatrica* 97: 1557-1563.

Popkin BM, Udry JR 1998. Adolescent obesity increases significantly in second and third generation U.S. immigrants, The National Longitudinal study of adolescent health. *J Nutr* 128: 701-706.

Raman RP 2002. Obesity and health risks. *J Am Coll Nutr* 21: 134S-139S.

Rogerson PA, Han D 2002. The effects of migration on the detection of geographic differences in disease risk. *Soc Sci Med* 55: 1817-1828.

Schneider D 2000. International trends in adolescent nutrition. *Soc Sci Med* 51: 955-967.

Shahar D, Shai I, Vardi H, Shahar A, Fraser D 2005. Diet and eating habits in high and low socioeconomic groups. *Nutrition* 21: 559-566.

Shang L, Xu Y, Jiang X, Hou R 2005. *Body mass index reference curves for children aged 0-18 years in Shaanxi, China*. IJBS 1: 57-66.

Shatenstein B, Ghadirian P 1998. Influences on Diet, health behaviours and their outcome in select ethnocultural and religious groups. *Nutrition* 14: 223-230.

Silventoinen K, Rokholm B, aprio J, Sorensen TIA 2010. *The genetic and environmental influences on childhood obesity: a systematic review of twin and adoption studies*. 34: 29-40.

Statistik Austria 2006. *Statistisches Jahrbuch Österreich Verlag Österreich, Austria*

Stellinga-Boelen AAM, Wiegersma PA, Bijleveld CMA, Verkade HJ 2007. Obesity in asylum seekers' children in the Netherlands – the use of national reference charts. *Eur J Pub Health* 17: 555-559.

Tanyolac S, Cikim SA, Azeli AD, Orhan Y 2008. Correlation between educational status and cardiovascular risk factors in a overweight and obese Turkish female population. *Anadolu Kardiyol Derg* 8: 336-341.

Tomkins A 2001. Nutrition and the health of Europe's children. *Public Health Nutr* 4: 89.

Toschke AM, Martin RM, von Kies R, Wells J, Smith GD, Ness AR 2007. Infant feeding method and obesity: body mass index and dual energy x ray absorptiometry measurements at 9-10 y of age from the Avon longitudinal study of parents and children (ALSPAC*). Am J Clin Nutr* 85: 1578-1585

Uitewaal PJM, Manna DR, Bruijnzeels MA, Hoes AW, Thomas S 2004. Prevalence of type 2 diabetes mellitus, other cardiovascular risk factors, and cardiovascular disease in Turkish and Moroccan immigrants in North West Europe: a systematic review. *Preventive Med* 39: 1068-1076.

Uitewaal PJM, Goudswaard AN, Ubnik-Veltmaat LJ, Bruijnzeels MA, Hoes AW, Siep T 2004. Cardiovascular risk factors in Turkish immigrants with type 2 diabetes mellitus: Comparison with Dutch patients. *Eur J Epidemiol* 19: 923-929.

Ujcic-Voortman JK, Schram MT, Jacobs-van der Bruggen MA, Verhoeff AP, Baan CA 2009. Diabetes prevalence and risk factors among ethnic minorities. *Eur J Public Health* 19: 511-515.

Ulijaszek SJ 2007. Obesity: a disorder of convenience. *Obes Reviews* 8 183-187.

Ulijaszek SJ, Lofink H 2006. *Obesity in Biocultural perspective*. Ann Rev Anthrop 35: 337-360.

UNECE 2006. *Recommendations for the 2010 census of Population and Housing*. United Nations Geneva

Vissandjee B, Desmeules M, Cao Z, Abdool S, Kazanjian A 2004. Integrating ethnicity and migration as determinants of Canadian women's health. *BMC Women's Health* 4: 32-43.

Wang Y, Wang JQ 2000. Standard definitions of child overweight and obesity worldwide. *Brit Med J* 321: 1158.

Weiss R, Kaufman FR 2008. Metabolic complications of childhood obesity. *Diabetes care* 31: 310-316.

Weiss R, Dziura J, Burget TS, Tamborlane WV, Taksali SE, Yeckel CW, Allen K, Lopes M, Savoye M, Morrison J, Sherwin RS, Caprio S 2004. Obesity and the metabolic syndrome in children and adolescents. *The New England J Med* 350: 2362-2374.

WHO 1998. *Obesity preventing and managing the global epidemic.* Geneva WHO

WHO 2006. Multicentre growth reference study group. WHO Child Growth Standards based on length/height, weight and age. *Acta Paediatrica Suppl* 450: 76-85

Will B, Zeeb H, Baune BT 2005. Overweight and obesity at school entry among migrant and German children: a cross-sectional study. *Public Health* 5: 45-52.

Wolin KY, Colangelo LA, Chiu BCH, Gapstur SM 2009. Obesity and immigration among Latina women. *J Immigrant Minority Health* 11: 428-431

Zarfl B, Elmadfa I 1995. Body Mass Index (BMI) als Indikator für Übergewicht bei Kindern und Jugendlichen – Ergebnisse der ASNS. *Aktuelle Ernährungsmedizin* 20: 201-206.

Zwiauer K, Wabitsch M. 1997. Relativer Body-mass-Index (BMI) zur Beurteilung von Übergewicht und Adipositas im Kindes- und Jugendalter. *Monatsschrift Kinderheilkunde* 145: 1312-1318.

In: Childhood Obesity: Risk Factors, Health Effects...
Editor: Carol M. Segel

ISBN: 978-1-61761-982-3
©2011 Nova Science Publishers, Inc.

Chapter VI

Promotion of Healthy Lifestyles to Prevent Obesity in School Children

Elena Centis, Rebecca Marzocchi and Giulio Marchesini[*]
University of Bologna, Italy

Abstract

All over the world obesity represents a social and medical emergency. The obesity epidemic is particularly alarming in young age, as overweight or obese children and teen-agers are likely to become obese adults. Overweight and obesity in childhood and in adolescence can be ascribed to genetic factors as well as to socio-environmental factors peculiar to the principal contexts of life. All over the world, public institutions are taking action to break this vicious circle. Also in Italy, national, regional and local projects have been developed to favor healthy lifestyles in children, following a prevention model integrating 'macro' actions (of political and social nature) with educational interventions directed to the 'micro' environment of individual subjects.

In this report, we revise the literature in the area of healthy life-style promotion in childhood, focusing on the results of a few Italian programs on the nutritional and physical habits of children. The main purpose was to demonstrate whether interventions aimed at increasing physical activity could produce significant changes in children's behavior, as well as awareness about the needs of healthy choices in the family.

1. Introduction

More than 50% of the adult Italian population is overweight or obese. The obesity epidemic is particularly alarming in children and in teen-agers. In Italy, 24% of children of the third grade of primary school are overweight and 12% are obese; in general, over a

---

[*] Address for correspondence: Prof. Giulio Marchesini, "Unit of Metabolic Diseases & Clinical Dietetics, "Alma Mater Studiorum" University of Bologna, 9, Via Massarenti, I-40138 Bologna, Italy

million children in the age range 6-11 have problems with obesity and overweight: more than one child in three. In over 60% of cases overweight/obesity develops before puberty and this makes the future bleak. There is indeed a high probability that overweight/obese children or teen-agers will become obese adults. The reasons for overweight and obesity in childhood and in adolescence are to be sought within individual genetic factors and life-styles as well as time trends of changes in socio-environmental factors, i.e., the characteristics of the main contexts of life of the individuals and of the community.

Overweight and obesity prevention in childhood and adolescence thus represents a priority target for public health. Concerted actions should be jointly carried out by the family, the school and those institutions that deal with health, nutrition and food supply, physical activity and transportation. Parents, teachers and, in general, every person responsible for children's education and health represent a model for correct nutritional behaviors and active lifestyle. Scientific literature recommends multi-disciplinary (multi-component) interventions applied to the school context, with the purpose of modifying the nutritional choices of children (e.g., promoting a larger consumption of fruit and vegetables), both in the short- and in the long-term period.

This review is aimed at providing a rationale for interventional planning on obesity prevention in children, underlining the importance of educational activities that can give a significant contribution to the promotion of a correct life-style.

2. Overweight and Obesity in Childhood and Adolescence: Dimensions and Causes

Childhood obesity is a major public health problem, given its increasing prevalence and adverse health consequences [1]. In the United States, the prevalence of obesity among children, defined as body mass index (BMI) higher than or equal to the 95th centile, has more than tripled since 1970, and the proportion of obese adolescents (12 to 19 years of age) increased from 4.6% in 1966–1970 to 17.4% in 2003–2004 [2, 3]. Similar trends have been observed in Canada [4], in the United Kingdom [5] and in Europe [6], involving all European countries to a similar extent. In Italy a significant increase of overweight and obesity has become manifest since the nineties, especially in the age-range between the 6 and 17 years. The updating of the HBSC (Health Behavior in School-aged Children) study showed that obesity is increasing, particularly in males and in the central-southern areas of Italy, in spite of a modest improvement in the consumption of fruit and vegetables and in the levels of physical activity in selected age ranges [7]. In addition to the growing numbers of obese children, the number of children with BMI higher than the 10th, 50th, 85th and 90th centile of norm continue to increase as well, in keeping with an increase in weight for height across the entire population [3, 8]. These trends are likely to result in a significant increase in the rates of coronary artery disease, hypertension, diabetes mellitus and other obesity-related diseases in young and middle-aged adults [9-11], and in an unprecedented decline in life expectancy in the developed world [12].

Social consequences should not be underestimated either: overweight or obese children are often teased by schoolmates and marginalized by school mates [13], which reduce their self-esteem with negative effects on future personality and character [14, 15].

2.1. Why are Overweight and Obesity Increasing in Childhood and in Adolescence?

According to a socio-ecological approach, the reasons for increased fat mass have to be sought within individual and environmental factors. Apart from genetic factors, the development of overweight and obesity in childhood and in adolescence has to be ascribed to the lifestyle change of modern society; the population is more sedentary and more inclined to consume high-energy, highly palatable foods, rich of sugar and salt. These unhealthy behaviors are often favored by current way of living, starting from the school age.

The pyramid in Figure 1 illustrates how obesity is determined by individual factors, with a limited effect, as well as by environmental factors, which produce much stronger effects at much larger levels [16]. Moving along the pyramid from top to bottom, several aspects become crucial in the "School" area. Nutritional habits and physical activity are behavioral aspects acquired during childhood and adolescence, but have an impact on future health *via* the development and growth of fat, fat-free and bone mass, as well as on the future prevalence of metabolic diseases, namely type 2 diabetes.

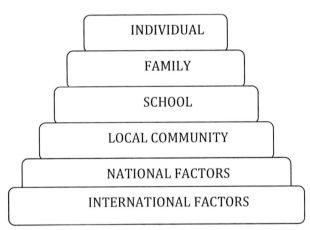

Note the different extension of factors, suggesting that intervention becomes more and more difficult from top to bottom, but the effects are larger when addressed to the general population

Figure 1. Causes of overweight and obesity in childhood and adolescence according to the model proposed by the International Obesity TaskForce.

The presence in the family of obese or overweight parents favors overweight or obesity in children. Children in the age range 5 to 10, either belonging to families with overweight or obese parents, or low-income families or with an obese and poorly-educated mother record a higher value of BMI since birth. In particular, the low educational level of the mother can promote incorrect nutritional behaviors and a sedentary lifestyle, as well as misperception of children's weight [17, 18]. The interventions of the Italian Ministry of Education regarding health promotion strategies (e.g., the possibility to practice physical activity at school, the quality of meals served at lunchtime) support correct nutritional habits and adequate levels of physical activity as the basis for prevention of overweight and obesity.

3. Prevention of Obesity at School

In the area of health promotion, the school is considered one of the most favorable settings to prevent overweight and obesity during growth. The majority of recent publications indicate the school setting as the main target area of intervention [19-21]. The school is the place where children and teen-agers spend most of the day; educational objectives include the implementation of healthy behaviors [22, 23]. The school could represent one of the most effective contexts to modify unhealthy habits: children in the age range 8 to 10 (end of primary school) already have food preferences and physical habits, but they still are susceptible to change if they receive positive reinforcement by parents, teachers, friends and schoolmates and are taught to analyze the models proposed by the media critically.

Interventions may be distinguished on the basis of their health targets [24] as interventions to reduce body weight (BMI) as against interventions to modify unhealthy behaviors (i.e., incorrect nutritional habits and sedentary life), which have the effects on body weight loss as a secondary target.

Interventions to reduce body weight show a conflicting or limited level of effectiveness. For this reason, the literature recommends interventions aimed at changing incorrect behaviors as pivotal actions to prevent overweight and obesity for all age ranges.

Table 1. Summary of interventions of proven efficacy in the treatment of childhood obesity

- multidisciplinary projects considering food education and/or promotion of physical activity in the school
- multidisciplinary projects focusing on food education
- interventions to increase the consumption of fruit and vegetables
- interventions aimed at changing/improving the type and frequency of snacks
- multidisciplinary projects promoting physical activity

The most effective interventions are summarized in Table 1 and detailed below.

- *Multidisciplinary projects considering food education and/or promotion of physical activity in the school.* These projects are based on changes in a few logistic and organizational characteristics of the school environment. Positive results have been obtained with such interventions both in the short- and in the long-term, especially in primary school children.
- *Multidisciplinary projects focusing on food education.* They are aimed at increasing the consumption of fruit and vegetables and/or to improve the management of school meals, including the way meals are prepared and the meal times.
- *Interventions to increase the consumption of fruit and vegetables.* They are particularly desirable in the primary school. These interventions get major positive and meaningful results when directed to children that consume a low number of portions of fruit and vegetables.

- *Interventions aimed at changing/improving the type and frequency of snacks.* They are suitable both in primary and in secondary school. In general, they are characterized by the increased availability of low-fat foods/snacks in the cafeteria and in the vending machines inside the schools.
- *Multidisciplinary projects promoting physical activity.* They are effective during the intervention period, but their long-term results are scarcely demonstrated. In primary school, they usually involve the school, as well as the family and the whole community. These projects consist in dynamic activities at school, an increase in the curricular time devoted to physical activity, the offer of extra-curricular sports activities, a limited time in front of TV or the computer at home, and are frequently supported by the use of a step-counter to monitor changes.

These different types of interventions are successful if the whole school community is involved and generated a positive mutual reinforcement. This is the primary purposes of the European Network for Health Promoting Schools.

An example of an effective intervention strategy involving the whole community has been adopted in France within the Epode (Ensamble Prevenons l'Obesitè des Enfants) project with the purpose of preventing childhood obesity by involving and making the whole population responsible for health changes.

The pilot study started in 1992 and involved two French cities. After a first phase, essentially based on nutritional educational interventions at school, the project was extended to the whole community. Gyms were built, sports trainers and dietitians were recruited, walks were organized to maintain physical activity in the leisure time, also involving children's parents. Nutritional education was developed, breakfast was distributed at school, cooking lessons were arranged for parents. Finally, in the local community chemists and food sellers were involved, supermarkets promoted the consumption of fruit by offering it at low prices and their car parks were made available after hours for events and sports activities. The general practitioners had a key role in children' weight and height surveillance and in the promotion of a culture about overweight and obesity, with the continuous support of media. Comparing the two seminal study towns with others bordering towns, it was shown that, after an initial trend towards a progressive increase of overweight in children, the prevalence of overweight declined (8.8% vs. 17.8% in control towns) by intensifying and extending the interventions to the whole community [25]. This project involved all components of the family. For example, the number of families that used to eat French fries once a week decreased from 56% to 39% and there was a stop to progressive weight increase in children's parents. The program has since then been adopted by almost 200 cities, not only in France, but also in Spain and Belgium and has recently moved to Greece and to extra-European countries with encouraging results.

In general, as also observed in different settings, the cumulative effect of small dietary changes may turn into a significant improvement in the whole population [26].

4. Nutritional and Physical Activity Habits of Primary School Children in Italy and in the Area of Bologna (Northern Italy)

A project has been set up at national level in Italy as an answer to the proposal by the World Health Organization to create a surveillance system on obesity in children. The project includes two phases; the first one, called "Okkio alla Salute"(Look at Health) [27], investigates the nutritional habits and the behavioral risk factors of primary school children. The survey was carried out in 2008 on a representative sample of 45.590 children attending the third class of primary school (8 years old). The second phase (2009) has adolescents as targets.

At national level, the prevalence of overweight and obesity was exceedingly high: 23.6% of children were overweight and 12.3% were obese, with a larger prevalence in Southern Italy.

The project collected a lot of data on the nutritional habits of children. In particular, 11% of children had no breakfast, 28% have breakfast but the nutritional intake is not healthy, 82% have a morning snack qualitatively incorrect, 23% of parents report that their children don't eat fruit and vegetables daily, only 2% of children eat more than the minimum recommended 4 portions/day of fruit and vegetables, 41% of children consume soft drinks every day, and 17% drink them more than once a day. Things are no better in the area of physical activity. Only 1 child in 10 complies with the recommended level of physical activity and 1 child out of 2 spends more than two hours/day watching television or in front of the computer playing video games. An elevated percentage of parents has a distorted perception of the nutritional status of their children: approximately 4 mothers in 10 of overweight/obese children do not think that their children have an excessive weight for height. This under-evaluation of the problem is larger in mothers with low educational level.

These results are perfectly in keeping with reports showing that universally parents are very likely to misperceive their children's weight. A distorted parental perception of childhood obesity represents a problem because it hinders the efforts to reduce body weight, the risk factors for pediatric obesity and its related complications [28].

The SAMBA project (SAMBA - Sorveglianza Attività Motoria Bambini, i.e, Physical Activity Surveillance in Children) is a triennial intervention (2006 -09) on 522 children (26 classes) of the third class randomized in Bologna. The project aims at the promotion of physical activity in children attending the third grades of primary school, with a follow up after 3 years. The first cross-sectional analysis showed a very high participation in sport practice (80.1% in the third class, 83.7% in the fifth class), but also a high prevalence of sedentary activities in leisure time (TV watching, 38.5% in the third class, 31.4% in the fifth class; Video games: 29.0% and 26.9%, respectively), a high prevalence of overweight (24.4%) and obesity (9.7%), and finally the lack of association between physical practice and BMI. In this age range the main factor regulating body weight is not the practice of sport, but daily habits, which are regulated by sedentary behaviors more than by leisure-time physical activity [29].

4.1. Interventional Study: "Misurare Il Cambiamento. Alimentazione E Attività Motoria Nella Scuola Primaria" (Measuring the Change. Nutritional and Physical Activity in the Primary School)

In keeping with the WHO recommendations and in order to promote health in the whole community, the project "Centro Salute G. C. Croce" was started in 2008 with the aim to sensitize a relatively small community to promote and maintain health starting from childhood obesity.

The project was developed along two interconnected lines: a) continuing education of General Practitioners and Pediatricians; b) an interventional activity in school children. The educational intervention was finalized to train GPs to a continuous support of the population in the area of healthy behaviors and correct lifestyles, with a specific attention to the principal risk factors for overweight and obesity. To facilitate the process of change, a session of the educational program was specifically devoted to motivational interviewing. The interventional program was addressed to primary school children. The primary goals were modifications of habits concerning nutrition and physical activity, as well as developing awareness in the family as to the needs of healthy choices.

The intervention was carried out in 210 children attending the fourth grade of the primary school in communities in the neighborhood of Bologna. They were allocated to an interventional (n = 104) and to a control arm (n = 106) in order to have a balanced composition as to socioeconomic status, presence of immigrants and parents' education. After several meetings with the governing boards of municipalities, school executives, teachers and the families, the height, weight, circumferences and triceps skinfolds of all children were measured and their habits in term of physical activity were recorded through questionnaire (filled by parents). The same measurements were repeated 6 to 8 months later, at the end of the intervention.

4.1.1. Intervention Activities and Results

The project planned to increase physical activity by 30 min/day, to be added to the curricular physical activity that children already practiced at school. This extra-activity was implemented with the support of experts with master in physical science, with specific competence in the practice of sports in young children. Once a week they had a meeting with the children and the teachers of the experimental group to implement new recreational motor activities to be performed during the following week both in the curricular and in the extra-curricular hours. Step counters were given to children to stimulate (and eventually measure) the daily amount of physical activity.

The nutritional intervention consisted in a meeting with teachers and children finalized to stimulate children to the correct nutritional practice of breakfast. The meeting, conducted by a physician and an expert of motivational activities, consisted in a brief theoretical lesson followed by practical lesson, in which every child worked with food to prepare a funny breakfast dish. The results of this part are reproduced in Figure 2.

Also the parents were invited to attend meetings, leaded by a physician expert in Nutritional Science on specific themes aimed at giving information and at sensitizing and motivating parents to practice healthier habits. In particular, during three different meetings, the parents were instructed on correct nutritional habits (the food pyramid), on advantages of physical activity, on the emotional aspects connected with food and promotion of correct food choices, and finally on recommendation for cooking.

Figure 2. Examples of the amusing breakfast dishes produced by children during the motivational lesson: Mexican "caballeros" – fruit & mint (left); clown – cereals, fruit & yogurt (center); tic-tac-toe – bread slices & jam (right)

In addition, the parents received weekly telephone calls by an expert of nutritional science. The purpose was to inform, sensitize and motivate parents to improve the life style of their children and to respect the project program. The telephone contacts with parents were also used to record data on ongoing changes in nutritional and physical habits of children (breakfast, snacks, fruit and vegetable consumption, meals composition, and physical activity). In addition, they provided an opportunity to reinforce education and offer a cognitive and educational support to parents and children, meet possible concerns and identify subjects with ongoing problems in adherence. Finally, they were used to monitor self-assessed weight, to be compared with final weight at study end.

The operators who made the phone-calls were chosen among the students of the postgraduate school of Nutritional Science and had been previously trained on the techniques of motivational interviewing. This part of the intervention is not new in the area of motivational activities. Telemedicine and home assistance through telephone have become common procedures in chronic diseases (mainly in old people), as well as in conditions where continuous care and supportive interventions are needed [30, 31]. Obesity is one of such conditions where the use of the telephone has emerged as a useful component in the promotion of healthy life styles and to maintain adherence to treatment [32, 33], as well as to target treatment to a more general audience than those attending specialized centers [34].

Finally, additional educational meetings, led by experts in Nutritional Science in collaboration with pediatricians, gave the teachers the competence to plan and set up educational interventions on food choices and nutrition during school time.

The results of the intervention are reported in Table 2. At entry into the study, no differences in anthropometric measurements were observed between the interventional and the control group. By the end of the program the body weight increased by over 2 kg in the control group and by only 1 kg in the intervention arm. Also height increased in both groups, but the different increase in body weight translated into significant differences in the changes in BMI, waist/hip ratio and in triceps skinfold thickness between groups. Although limited follow-up is so far available, the results indicate that the intervention was able to modify the time trend of progressive increase in BMI observed in childhood. Notably, we observed an enthusiastic participation of children in motor activities and in food preparation, when these were proposed as joyful activities.

Table 2. Anthropometric measurements in children in the experimental and in the control arm (means ± SD).

Measurements	Control group baseline	Follow-up	Intervention group baseline	Follow-up	P*
Weight (kg)	34.8 ± 7.3	37.1 ± 7.8	33.2 ± 7.9	34.3 ± 8.0	<0.001
Height (cm)	1.37 ± 0.6	1.39 ± 0.7	1.36 ± 0.6	1.37 ± 0.7	NS
BMI (kg/m^2)	18.5 ± 3.0	19.1 ± 3.1	17.9 ± 3.2	18.0 ± 3.1	<0.001
Waist circumference (cm)	67.7 ± 8.4	69.5 ± 8.7	65.0 ± 8.9	66.3 ± 8.7	0.439
Hip circumference (cm)	63.0 ± 6.9	63.6 ± 6.8	60.6 ± 6.7	61.4 ± 6.5	0.208
Waist/hip ratio	1.07 ± 0.05	1.09 ± 0.05	1.07 ± 0.05	1.08 ± 0.05	0.049
Triceps skinfold thickness (mm)	14.7 ± 5.1	16.2 ± 5.4	14.3 ± 5.6	14.3 ± 4.9	0.041
Arm circumference (cm)	21.1 ± 2.8	21.8 ± 2.8	20.4 ± 2.9	20.8 ± 2.6	0.208

*Data were tested for significance as repeated measures, time x treatment ANOVA

5. Conclusions

Multi-component interventions require remarkable care (not only in terms of time) from the project group and the intermediary recipients, represented by teachers and parents. This often represents a critical problem in the set up of obesity prevention projects. An effective involvement and collaboration from the teaching staff can be reached by long-term educational programs, carried out during the whole length of the intervention. These programs give participants the possibility of planning and experimenting behavioral models, methods and validated tools.

The involvement of families is often time-consuming and difficult to pursue, but it is the pivotal factor to give continuity to the educational intervention at school and to reinforce correct lifestyles in children. It is important to encourage communication and exchanges among members of the steering group, teachers and parents, to get a feedback on feelings, needs and operative ideas about new activities to be proposed at school, to give informative material with practical advice to support the messages conveyed at school and involve parents in the same educational activities as their children. Only a comprehensive intervention involving all actors of children's education is expected to generate long-lasting effects.

References

[1]   Reilly JJ, Methven E, McDowell ZC, Hacking B, Alexander D, Stewart L, et al. Health consequences of obesity. *Arch Dis Child* 2003;88:748-752.
[2]   Ogden CL, Carroll MD, Curtin LR, McDowell MA, Tabak CJ, Flegal KM. Prevalence of overweight and obesity in the United States, 1999-2004. *JAMA* 2006;295:1549-1555.
[3]   Ogden CL, Flegal KM, Carroll MD, Johnson CL. Prevalence and trends in overweight among US children and adolescents, 1999-2000. *JAMA* 2002;288:1728-1732.

[4] Shields M, Tremblay MS. Canadian childhood obesity estimates based on WHO, IOTF and CDC cut-points. *Int J Pediatr Obes* 2010;5:265-273.

[5] Jotangia D, Moody A, Stamatakis E. *Obesity among children under 11*. London: UK Department of Health; 2006.

[6] Rolland-Cachera MF, Castetbon K, Arnault N, Bellisle F, Romano MC, Lehingue Y, et al. Body mass index in 7-9-y-old French children: frequency of obesity, overweight and thinness. *Int J Obes Relat Metab Disord* 2002;26:1610-1616.

[7] Currie C, Nic Gabhainn S, Godeau E. The Health Behaviour in School-aged Children: WHO Collaborative Cross-National (HBSC) study: origins, concept, history and development 1982-2008. *Int J Public Health* 2009;54 Suppl 2:131-139.

[8] Fredriks AM, van Buuren S, Wit JM, Verloove-Vanhorick SP. Body index measurements in 1996-7 compared with 1980. *Arch Dis Child* 2000;82:107-112.

[9] Bibbins-Domingo K, Coxson P, Pletcher MJ, Lightwood J, Goldman L. Adolescent overweight and future adult coronary heart disease. *N Engl J Med* 2007;357:2371-2379.

[10] Hayman LL, Williams CL, Daniels SR, Steinberger J, Paridon S, Dennison BA, et al. Cardiovascular health promotion in the schools: a statement for health and education professionals and child health advocates from the Committee on Atherosclerosis, Hypertension, and Obesity in Youth (AHOY) of the Council on Cardiovascular Disease in the Young, American Heart Association. *Circulation* 2004;110:2266-2275.

[11] Sorof JM, Lai D, Turner J, Poffenbarger T, Portman RJ. Overweight, ethnicity, and the prevalence of hypertension in school-aged children. *Pediatrics* 2004;113:475-482.

[12] Olshansky SJ, Passaro DJ, Hershow RC, Layden J, Carnes BA, Brody J, et al. A potential decline in life expectancy in the United States in the 21st century. *N Engl J Med* 2005;352:1138-1145.

[13] Griffiths LJ, Page AS. The impact of weight-related victimization on peer relationships: the female adolescent perspective. *Obesity* (Silver Spring) 2008;16 Suppl 2:S39-45.

[14] Must A, Strauss RS. Risks and consequences of childhood and adolescent obesity. *Int J Obes Relat Metab Disord* 1999;23 Suppl 2:S2-11.

[15] Puhl RM, Latner JD. Stigma, obesity, and the health of the nation's children. *Psychol Bull* 2007;133:557-580.

[16] Finkelstein E, French S, Variyam JN, Haines PS. Pros and cons of proposed interventions to promote healthy eating. *Am J Prev Med* 2004;27:163-171.

[17] Manios Y, Kondaki K, Kourlaba G, Vasilopoulou E, Grammatikaki E. Maternal perceptions of their child's weight status: the GENESIS study. *Public Health Nutr* 2009;12:1099-1105.

[18] Guendelman S, Fernald LC, Neufeld LM, Fuentes-Afflick E. Maternal perceptions of early childhood ideal body weight differ among Mexican-origin mothers residing in Mexico compared to California. *J Am Diet Assoc* 2010;110:222-229.

[19] de Silva-Sanigorski A, Prosser L, Carpenter L, Honisett S, Gibbs L, Moodie M, et al. Evaluation of the childhood obesity prevention program Kids--'Go for your life'. *BMC Public Health* 2010;10:288.

[20] Della Torre Swiss SB, Akre C, Suris JC. Obesity prevention opinions of school stakeholders: a qualitative study. *J Sch Health* 2010;80:233-239.

[21] Brownson RC, Chriqui JF, Burgeson CR, Fisher MC, Ness RB. Translating epidemiology into policy to prevent childhood obesity: the case for promoting physical activity in school settings. *Ann Epidemiol* 2010;20:436-444.

[22] Doak CM, Visscher TL, Renders CM, Seidell JC. The prevention of overweight and obesity in children and adolescents: a review of interventions and programmes. *Obes Rev* 2006;7:111-136.

[23] Sharma M. International school-based interventions for preventing obesity in children. *Obes Rev* 2007;8:155-167.

[24] Summerbell CD, Waters E, Edmunds LD, Kelly S, Brown T, Campbell KJ. *Interventions for preventing obesity in children: Cochrane Database of Systematic Review Art.* No. CD001871; 2005.

[25] Romon M, Lommez A, Tafflet M, Basdevant A, Oppert JM, Bresson JL, et al. Downward trends in the prevalence of childhood overweight in the setting of 12-year school- and community-based programmes. *Public Health Nutr* 2009;12:1735-1742.

[26] Paineau D, Beaufils F, Boulier A, Cassuto DA, Chwalow J, Combris P, et al. The cumulative effect of small dietary changes may significantly improve nutritional intakes in free-living children and adults. *Eur J Clin Nutr* 2010;(n press).

[27] Spinelli A, Lamberti A, Baglio G, Andreozzi S, Galeone D. Okkio alla Salute: sistema di sorveglianza su alimentazione e attività fisica nei bambini della scuola primaria: *Rapporti ISTISAN* 09/24; 2009.

[28] Doolen J, Alpert PT, Miller SK. Parental disconnect between perceived and actual weight status of children: a metasynthesis of the current research. *J Am Acad Nurse Pract* 2009;21:160-166.

[29] Leoni E, Beltrami P, Poletti G, Baldi E, Sacchetti R, Garulli A, et al. [Survey on sports practice and physical activity of primary school children living in the area of Bologna Local Health Unit in relation with some individual and environmental variables]. *Ann Ig* 2008;20:441-453.

[30] Siegel K, Mesagno FP, Karus DG, Christ G. Reducing the prevalence of unmet needs for concrete services of patients with cancer. Evaluation of a computerized telephone outreach system. *Cancer* 1992;69:1873-1883.

[31] Woollard J, Beilin L, Lord T, Puddey I, MacAdam D, Rouse I. A controlled trial of nurse counselling on lifestyle change for hypertensives treated in general practice: preliminary results. *Clin Exp Pharmacol Physiol* 1995;22:466-468.

[32] Brekke HK, Jansson PA, Lenner RA. Long-term (1- and 2-year) effects of lifestyle intervention in type 2 diabetes relatives. *Diabetes Res Clin Pract* 2005;70:225-234.

[33] Byrne SM, Cooper Z, Fairburn CG. Psychological predictors of weight regain in obesity. *Behav Res Ther* 2004;42:1341-1356.

[34] Jeffery RW, Sherwood NE, Brelje K, Pronk NP, Boyle R, Boucher JL, et al. Mail and phone interventions for weight loss in a managed-care setting: Weigh-To-Be one-year outcomes. *Int J Obes Relat Metab Disord* 2003;27:1584-1592.

In: Childhood Obesity: Risk Factors, Health Effects...   ISBN: 978-1-61761-982-3
Editor: Carol M. Segel   ©2011 Nova Science Publishers, Inc.

Chapter VII

Using Real-Time Data Capture Methods to Investigate Children's Physical Activity and Eating Behaviors

Genevieve Fridlund Dunton
University of Southern California, CA, USA

Abstract

Advances in portable electronic technologies have created opportunities for the real-time assessment of children's physical activity and eating behaviors in naturalistic situations. Mobile phones or PDA's can be used to record electronic surveys, take photographs, or indicate geographic locations of children's behaviors. Unlike self-report instruments, which are prone to recall errors and biases, real-time data capture (RTDC) methods can assess behaviors as they occur. In addition, these strategies are able to provide contextual information about physical activity and healthy eating such as where and with whom the behaviors are taking place; and how children feel before, during, and after these activities. This commentary will describe how RTDC methods can enhance our understanding of factors influencing children's physical activity and eating behaviors. In particular, it will discuss the potential to advance research pertaining to the following questions: (1) How frequently, when, what amount, what intensity, what duration, and what type of food or activity was eaten or performed? (2) Where and with whom do children engage in physical activity and eat, and do these patterns differ according to demographic (e.g., sex, age, ethnic, income) and temporal (e.g., time of day, day of the week, seasonal) characteristics?; (3) How do children's physical activity levels (e.g., intensity, duration) and eating patterns (e.g., amount/content of food) differ across physical and/or social contexts?; (4) To what extent do mood, stress, and psychosocial factors serve as time-related antecedents and consequences to children's physical activity and eating episodes?; and (5) Are patterns of within-daily variability in children's physical activity and eating behaviors related to health outcomes such as body weight, insulin dependence, and the metabolic syndrome? The commentary will also discuss practical and economic challenges associated with employing RTDC methodologies in research studies with children. It will conclude by addressing how these innovative research strategies can inform the design of programs and policies to prevent and treat childhood obesity.

Introduction

Rising rates of overweight and obesity among U.S. children are leading to concerns about the increased risk of diabetes, sleep apnea, hypertension, and other cardiovascular and metabolic disorders across the life course [1-4]. Obesity is largely thought to result from an imbalance between energy intake (i.e., dietary patterns) and expenditure (i.e., physical activity). Recent estimates suggest that 34% of U.S. high school students consume soda or pop at least one time per day and only 21% consume fruits and vegetables five or more times per day [5]. On the side of energy expenditure, only 40-50% of 6-11 year-old, 6-11% of 12-15 year-old, and 7.6% of 16-19 year-old youth engage in ≥60 minutes per day of moderate-intensity activity on at least 5 out of the past 7 days [6]. Also, 6-19 year-old children spend an average of 5-8 hours per day in sedentary activity [7]. Since health behavior patterns practiced by youth have been shown to continue into adulthood [8,9] establishing healthy physical activity habits during childhood is a critical public health concern.

Advantages of Real-Time Data Capture to Assess Eating and Physical Activity

Surveillance, epidemiological, and intervention studies seeking to promote healthy eating and physical activity among youth rely upon informative methods of measuring these behaviors and their correlates. Precise assessment strategies are necessary to understand the dose-response relationship with various health outcomes, assess associations with environmental and psychosocial factors, and evaluate the effect of interventions and policies [10,11]. However, developing reliable and valid self-report methods of measuring children's eating and activity behaviors is challenging due to recall errors and biases [12-14]. Children often experience difficulties remembering the types and amount of foods consumed and duration of activities after 24 hours or more has passed since the behavior [15]. Although objective methods of assessing physical activity are available such as accelerometers and pedometers, these devices are unable to provide information about activity type, mood, or context. Many of these limitations can be overcome through real-time data capture (RTDC) methodologies.

Advances in portable electronic technologies have created opportunities for the real-time assessment of children's physical activity and eating behaviors in naturalistic situations. Mobile phones or PDA's can be used to record electronic surveys, take photographs, or indicate geographic locations of children's behaviors [16]. Unlike self-report instruments, which are prone to recall errors and biases, RTDC methods can assess behaviors as they occur. In addition, these strategies are able to provide contextual information about physical activity and healthy eating such as where and with whom the behaviors are taking place; and how children feel before, during, and after these activities [17,18]. Experience Sampling Methods (ESM) [19] and Ecological Momentary Assessment (EMA) [20] elicit electronic survey responses in real time. In event-contingent sampling, participants record information during or after a pre-determined behavior such as a meal or bout of physical activity. Interval-contingent sampling triggers survey responses or other types of data collection (e.g., photographs) according to a specific pre-set time frames (e.g., at 8am and 12noon everyday). Lastly, signal-contingent sampling schemes require participants to record data whenever they are prompted by the device, often at random times throughout the day [21,22].

Potential for RTDC Methodology to Inform Research in Key Areas

RTDC methodologies have the potential to enhance our understanding of children's eating and physical activity behaviors in a number of key areas. First, they can measure how frequently, with what intensity, what is taken in, how long, and when (e.g., day of week, time of day, season of the year) children engage in eating and physical activity. Second, they can assess where and with whom children engage in physical activity and eat (and how do they feel during these behaviors), in addition to whether these contextual patterns differ according to demographic (e.g., sex, age, ethnic, income) and temporal (e.g., time of day, day of the week, seasonal) characteristics. Third, RTDC methodologies can be used to examine whether children's physical activity levels (e.g., intensity, duration) and eating patterns (e.g., amount/content of food, duration of episodes) and mood during these behaviors differ across physical and/or social contexts. Fourth, they can evaluate the extent to which mood, stress, and psychosocial factors serve as time-related antecedents and consequences to children's physical activity and eating episodes. Lastly, RTDC strategies can determine whether patterns of day-to-day and within-daily variability in children's mood, stress, and psychosocial factors are related to physical activity and eating behaviors; and whether patterns of day-to-day variability and within-daily variability physical activity and eating behaviors are associated with negative health outcomes such as body weight, insulin dependence, and the metabolic syndrome? The key areas of research that can be informed by RTDC methodologies are described in further detail below.

Area 1: To Describe the Prevalence of Eating and Physical Activity Behaviors

RTDC methods can use electronic surveys, cameras, and Global Positioning Systems (GPS) devices to describe the prevalence of eating and physical activity behaviors among subgroups of children. These types of studies measure behavioral dimensions such as how frequently, when, what amount, what intensity, what duration, and what type of food or activity was eaten or performed. For instance, Dunton and colleagues [23] used a signal-contingent sampling scheme (every 30 +/- 10 minutes across 4 days) to measure the prevalence of walking and sports/exercise using electronic diaries in high schools students. This study found that the typical duration of exercise sessions was longer than walking sessions. Also, significantly higher rates of walking and exercise occurred on weekdays as compared with weekend days. In a study of eating behaviors among children ages 8-13 years,

RTDC methods were used to assess the type and quality of food and drinks consumed. Four days of signal- and event- contingent sampling conducted through interviews on child-specific cell phones found that episodes where children reported a loss of control over eating had greater intake of calories and carbohydrates than during regular meals [24]. These studies offer examples of RTDC strategies can assess the occurrence and patterning of eating and activity behaviors in naturalistic settings without the need for multiple-day recall.

Area 2: To Describe the Contexts of Eating and Physical Activity Behavior

Another area of research where RTDC methods are useful is the description of the physical and social contexts where children's eating and physical activity takes place. A growing body of evidence suggests that environmental factors such as the availability of parks, food outlets, and social support may play a role in shaping children's energy balance-related behaviors [25,26]. However, information is lacking on the extent to which children *actually visit* or *use* these contexts. Event- or signal-contingent ESM can assess where (e.g., home, school, park, restaurant) and with whom (e.g., friends, parents, siblings) a behavior occurs. In a recent electronic diary study of high school students, Dunton and colleagues [27] found that boys were more likely to report exercising and walking in outdoor locations than girls. Also, exercising with classmates, family, and at school decreased as adolescents progressed across high school. Students reported a greater proportion of their exercise and walking at school on weekdays compared with weekends. Furthermore, electronic diaries revealed that high school students were more likely to report exercising and walking outdoors in the fall and the spring than in the winter. Other RTDC methodologies such as GPS can capture the exact geographical location of that behavior using longitudinal and latitudinal coordinates [28,29]. For example, Wiehe and colleagues (2008) used GPS-enabled cell phones to track adolescent travel patterns. They found that adolescents' travel paths frequently extended outside of their immediate neighborhoods and there was substantial day-to-day variability in distance from home and direction of travel. To date, few studies have used RTDC strategies to describe children's eating contexts, although this is an intriguing area for further investigation.

Area 3: To Examine Differences in Eating and Physical Activity Across Contexts

In addition to telling us about where and with whom children's eating and physical activity takes place, RTDC methods can be used to examine differences in the level, type, and content of these behaviors across contexts. To date, little is known about how environmental characteristics of a setting impact the intensity of activity or type of food consumed by children in those situations. Behavior setting theory suggests that characteristics of the immediate context can shape mood and behavior occurring in that environment [31]. Electronic diary EMA studies have shown that children are more motivated and engage in higher intensity activities when in the presence of others as compared to when they are alone [32,33]. Also, the total number of steps and moderate-to vigorous physical activity were greater outdoors than at home or at someone else's house in a recent study EMA study in children using mobile phone surveys [34]. Identifying which types of physical and social environments are conducive to healthy eating and physical activity through RTDC methods can inform the development of context-specific policy and programmatic efforts in these areas.

Area 4: To Investigate Antecedents and Consequences of Eating and Physical Activity Episodes

RTDC strategies also have the ability to allow us to investigate temporal antecedents and consequences of eating and physical activity episodes in children. ESM and EMA can be conducted in a time-intensive manner such that there are multiple records or observations made across each day of monitoring. Individual observations may be separated by as little as 30 minutes or up to 4 or more hours. These interval- and signal contingent sampling schemes allow researchers to examine the time-lagged effects of mood, stress, social interactions, and other psychosocial factors on subsequent eating and physical activity episodes; as well as the effects of eating and physical activity on subsequent experiences and events. For example, in the cell phone interview EMA study conducted by Hilbert and colleagues [24], episodes of eating where children reported losing control were preceded and followed by thoughts about food, eating, dieting, and body image. However, negative moods states did not serve as precursors to loss of control eating episodes. Antecedents and consequences of physical activity episodes can also studies using RTDC methods. Although research on this topic is lacking in children, an EMA study of middle-aged and older adults found that physical activity had a positive effect on subsequent mood valence, arousal, and calmness [35]. Overall, RTDC strategies offer a promising opportunity for us to understand how children's feelings, events, experiences, and behaviors may trigger, exacerbate, inhibit, or change as a result of eating and physical activity episodes occurring across the day.

Area 5: To Examine Daily and Within-Daily Variability in Eating and Physical Activity

An understudied research area that could be greatly informed by RTDC methods is the extent to which daily and within-daily variability in eating and physical activity impact short- and long-term health outcomes in children. Observational and epidemiological studies of eating and physical activity typically evaluate the associations between the mean levels of target variables (e.g., daily minutes of moderate-to-vigorous physical activity) and health risks. However, there is growing interest in understanding whether the degree of daily and within-day fluctuations (i.e., variations from the person-level mean) in eating and physical activity can influence key health indicators such as body weight, insulin dependence, and the metabolic syndrome. If daily and within-daily variations are important, it would be helpful to know what types of factors predict these fluctuations (e.g., daily schedule, time demands, weather) and whether this variability changes (increases or decreases) over time for children, especially as they mature through puberty. This intraindividual approach to understanding human behavior has gained attention recent years as technological and statistical advancements allow for more fine-grained assessments and analysis of these within-person processes [36,37].

Practical and Financial Challenges of RTDC

Although RTDC methods have become increasing accessible in recent years, there are a number of practical and financial obstacles to using them in children's eating and physical activity research. First, they are vulnerable to problems with missing and ambiguous data, which complicate the data reduction and analysis process. Data may be missing for a variety of reasons, among them (1) participants forget to wear or carry the electronic devices, (2) participants are unable to or do not want to respond to device-initiated surveys when prompted, and (3) device limitations such as low battery life and malfunction. Event-contingent sampling that relies upon self-initiated electronic surveying after particular activities (e.g., exercise episodes) is also prone to various types of missing and ambiguous data resulting from delayed reports and the failure to report events altogether. Nonresponse rates vary across studies. In an EMA study using real-time mobile phone-based interview to assess social context, interest in activities, and positive and negative mood related to physical activity among children ages 10-17 years, 10% of the data were missing due missed calls [17]. A similar study using a phone-call based EMA strategy in 11-19 year-olds had a 36% missed call rate [18]. Research performed by Dunton and colleagues [38], which used signal-contingent electronic phone surveys to measure physical activity, sedentary behavior, mood, and contextual characteristics yield a nonreponse rate of 23%. Although most of these studies have a compensation structure based upon number of responses given, future studies would benefit from developing EMA protocols and/or items that are intrinsically rewarding and fun for children to answer, or use game-like strategies to promote compliance.

A second set of challenges for EMA studies of children's physical activity and eating behaviors involves reactance to items and the potential for participant burden. Although the objective of most EMA studies is to observe children's behavior without influencing it, the repetitive exposure to items pertaining to physical activity and eating may trigger children to think about or behavior in ways that they otherwise wouldn't. In fact, there is some evidence to suggest that the mere act of measuring a behavior could have some impact on that behavior in the future [39]. Reactance to EMA was examined in a study using electronic diaries to assess patterns of smoking in adults [40]. They found no difference in smoking abstinence between the EMA and no EMA groups. However, some scales such as motivation, temptation differed between groups suggesting reactivity to EMA. Participant burden can also limit the quality and quantity of data collected in EMA research. If the rate of prompting is too high and questions are too repetitive, participants may opt not to respond to the surveys or drop out of the study altogether. Other problems include the mindless answering of items, choosing the first response option for every item, or handing the device off to friend or family member to complete the survey because the respondent has become bored and/or want to finish the survey faster. To reduce concerns about participant reactance and burden, researchers should aim to use the fewest number of prompted surveys possible to answer their question or interest. For instance, if they are interested in assessing a relatively rare behavior such as a bout of vigorous exercise, they could choose to use an event-contingent measurement schedule instead of interval-contingent schedule with frequent sampling in hopes of catching the behavior. Also, researchers may choose to assess a random subset of items during each prompt.

A third challenge associated with the use of real-time data capture strategies to assess children's physical activity and eating patterns is the financial cost. Often an experienced computer programmer and several rounds of piloting testing are needed to develop EMA programs to run on mobile phones or PDA's. Also, the mobile phone devices themselves can be a costly research expense. However, software can be written to run on common mobile phones that are used by older children and adolescents such that the program can be loaded directly onto their current phone or the participant can temporarily swap their SIM card into a study phone so that they may make and receive personal calls on it during the study. Free open source EMA programs such as MyExperience (http://myexperience.sourceforge.net/) are also available and can be tailored to the researchers' specifications. Lastly, the number of and training necessary for research personnel to oversee EMA studies can cause some financial burden. Staff members often need to be available on a 24-hour basis to respond to and solve technical concerns, which may involve bringing a new device to the participants to replace the malfunctioning one. Personnel time is are also necessary to conduct phone calls to participants during the monitoring period to encourage compliance. However, with improvements in technology and the ability to develop systems with wifi-enabled transmission of EMA data to central servers, compliance can be monitored in real time to trigger automated reminder texts and cues.

Potential of RTDC to Inform Obesity Prevention Programs and Policies

EMA studies of children's physical activity and eating have to potential to inform the development of obesity prevention programs and policies. First, by knowing with whom and where children eat unhealthy foods and engaging in sedentary pursuits, researchers can construct context-specific interventions aimed and curbing obesogenic behaviors in particularly risky environments. Second, information about how children's moods and motivational factors can trigger eating certain foods or participating in certain activities can be helpful for parents and teachers to recognize those cues ahead of time and interfere with those behavioral process in a proactive manner. Third, findings from EMA studies can lead to the development of Ecological Momentary Interventions (EMIs), which use real-time prompting, questioning, and cueing to influence problem-solving, decision-making process, and behaviors related to obesity [16,41]. In a randomized controlled study, Hurling and colleagues [42] used internet and mobile phone technology to deliver an automated physical activity program to middle-aged adults. The intervention group tailored received text-based automated recommendations to overcome perceived barriers, a schedule to plan activity sessions, and instantaneous feedback about activity levels. After 3 months, the intervention group increased physical activity to a greater extent and lost more percent body fat than the control group. Also, an automated test-message intervention for college students that delivered personalized messages about health eating and weight loss two to five times daily resulted in greater weight loss than a comparison group after 4 months [43]. These studies offer initial support for the translation of evidence gathered through real-time data capture methods into electronically-delivered interventions to modify behaviors as they occur in naturalistic settings.

Conclusion

Advances in portable electronic technologies have created opportunities for the real-time assessment of children's physical activity and eating behaviors in naturalistic situations. Mobile phones or PDA's can be used to record electronic surveys, take photographs, or indicate geographic locations of children's behaviors. These strategies allow researchers to gather information that isn't readily available through recall instruments or accelerometers such as where, with whom, and how children feel before, during and after eating and activity episodes. RTDC methods have some challenges such as larger proportions of missing data than traditional instruments, participant reactance and burden, and greater initial research costs. However, the nature and quality of the data they provide has the potential to greatly inform obesity prevention programs and policies for youth, especially those interventions that are delivered in real-time using portable electronic devices.

References

[1]   Alisi A; Manco M; Panera N; Nobili V. Association between type two diabetes and non-alcoholic fatty liver disease in youth. *Annals of Hepatology,* 2009 8, 44-50.
[2]   Maclaren NK; Gujral S; Ten S; Motagheti R. Childhood obesity and insulin resistance. *Cell Biochemistry and Biophysics*, 2007 48, 73-8.
[3]   Arens R; Muzumdar H.J. Childhood obesity *Journal of Applied Physiology*, 2010 108, 436-44.
[4]   Virdis A; Ghiadoni L; Masi S; Versari D; Daghini E; Giannarelli C; Salvetti A; Taddei S. Obesity in the childhood: a link to adult hypertension. *Current Pharmaceutical Design,* 2009 15, 1063-71.
[5]   Centers for Disease Control and Prevention. Youth Risk Behavior Surveillance — United States, 2007. *Morbidity and Mortality Weekly Report*, 2008 57, SS-4.
[6]   Troiano RP; Berrigan D; Dodd KW; Masse LC; Tilert T; McDowell M. Physical activity in the United States measured by accelerometer. *Medicine and Science in Sports and Exercise,* 2008 40, 181-88.
[7]   Whitt-Glover MC; Taylor WC; Floyd MF; Yore MM; Yancey AK; Matthews CE. Disparities in physical activity and sedentary behaviors among US children and adolescents: prevalence, correlates, and intervention implications. *Journal Public Health Policy,* 2009 30,309-34.
[8]   Kelder SH; Perry CL; Klepp KI; Lytle LL. Longitudinal tracking of adolescent smoking, physical activity, and food choice behaviors. *American Journal of Public Health*, 1994 84, 1121-26.

[9]   Telama R; Yang X; Viikari J; Välimäki I; Wanne O; Raitakari O. Physical activity from childhood to adulthood: A 21-year tracking study. *American Journal of Public Health,* 2005 28, 267-73.

[10] Corder K; Ekelund U; Steele RM; Wareham NJ; Brage S. Assessment of physical activity in youth. *Journal of Applied Physiology,* 2008 105, 977-87.

[11] Welk; Corbin CB; Dale D. Measurement issues in the assessment of physical activity in children. Research Quarterly for Exercise Sport, 2000 71, 59-73.

[12] Sirard JR; Pate RR. Physical activity assessment in children and adolescents. *Sports Medicine, 2001 31,439-54.*

[13] Going SB; Levin S; Harrell J; Stewart D; Kushi L; Cornell CE; Hunsberger S; Corbin C; Sallis J. Physical activity assessment in American Indian schoolchildren in the Pathways study. *American Journal of Clinical Nutrition,* 1999 69,788-95.

[14] Sallis JF. Self-report measures of children's physical activity. *Journal of School Health,* 1991 61,215-9.

[15] Baxter SD. Cognitive processes in children's dietary recalls: insight from methodological studies. *European Journal of Clinical Nutrition,* 2009 63,19-32.

[16] Patrick K; Griswold WG; Raab F; Intille SS. Health and the mobile phone. *American Journal of Preventive Medicine,* 2008 35, 177-81.

[17] Axelson DA; Bertocci MA; Lewin DS; Trubnick LS; Birmaher B; Williamson DE; Ryan ND; Dahl RE. Measuring mood and complex behavior in natural environments: use of ecological momentary assessment *Journal of Child Adolescent Psychopharmacology,* 2003 13, 253-66.

[18] Rofey DL; Hull EE; Phillips J; Vogt K; Silk JS; Dahl RE. Utilizing ecological momentary assessment in pediatric obesity to quantify behavior, emotion, and sleep. *Obesity,* 2010 18,1270-2.

[19] Csikszentmihalyhi M; Larson R. Validity and reliability of the Experience-Sampling Method. *Journal of Nervous and Mental Disorders,* 1987 175, 526-36.

[20] Stone AA; Shiffman S. Ecological momentary assessment in behavioral medicine. *Annals of Behavioral Medicine,* 1994 16,199-202.

[21] Wheeler L; Reis HT. Self-recording of everyday life events: origins, types, and uses. *Journal of Personality,* 1991 59, 339-54.

[22] Reis, H.T. and Gable, S. Event-sampling and other methods for studying everyday experience. In H.T. Reis and C. M. Judd. *Handbook of Research methods in Social and Personality Psychology.* Cambridge, UK. Cambridge University Press; 2000;190-222.

[23] Dunton GF; Whalen CK; Jamner LD; Henker B; Floro JN. Using ecologic momentary assessment to measure physical activity during adolescence. *American Journal of Preventive Medicine,* 2005 29, 281-7.

[24] Hilbert A; Rief W; Tuschen-Caffier B; de Zwaan M; Czaja J. Loss of control eating and psychological maintenance in children: an ecological momentary assessment study. *Behavior Research and Therapy,* 2009 47, 26-33.

[25] Salmon J; Timperio A. Prevalence, trends and environmental influences *Medicine and Sport Science,* 2007 50,183-99.

[26] Sallis JF; Glanz K. The role of built environments in physical activity *The Future of Children,* 2006 16, 89-108.

[27] Dunton GF; Whalen CK; Jamner LD; Floro JN. Mapping the social and physical contexts of physical activity across adolescence using ecological momentary assessment. *Annals of Behavioral Medicine,* 2007 34, 144-53.

[28] Duncan MJ; Badland HM; Mummery WK. Applying GPS to enhance understanding of transport-related physical activity. *Journal of Science and Medicine in Sport,* 2009 12, 549.

[29] Stopher P; FitzGerald C; Zhang J. Search for a global positioning system device to measure person travel. *Transportation Research - Part C Emerging Commercial Tecnologies,* 2008 16, 350-69.

[30] Wiehe SE; Carroll AE; Liu GC; Haberkorn KL; Hoch SC; Wilson JS; Fortenberry JD. Using GPS *International Journal of Health Geographics,* 2008 7,22.

[31] Barker RG. *Ecological Psychology: Concepts and methods for studying the environment of human behavior.* Palo Alto, CA: Stanford University Press; 1968.

[32] Salvy SJ; Roemmich JN; Bowker JC; Romero ND; Stadler PJ; Epstein LH. Effect of peers and friends on youth physical activity and motivation to be physically active. *Journal of Pediatric Psychology,* 2008 34,217-25.

[33] Salvy SJ; Bowker JW; Roemmich JN; Romero N; Kieffer E; Paluch R; Epstein LH. Peer influence on children's physical activity: an experience sampling study. *Journal of Pediatric Psychology,* 2008 33, 39-49.

[34] Dunton GF; Liao Y; Intille S; Wolch J; Pentz M. Social and physical contextual influences on children's physical activity: An ecological momentary assessment study. *Journal of Physical Activity and Health,* in press.

[35] Kanning M; Schlicht W. Be active and become happy: an ecological momentary assessment of physical activity and mood. *Journal of Sport and Exercise Psychology,* 2010 32, 253-61.

[36] Cervone D. Personality architecture: within-person structures and processes. *Annual Review of Psychology,* 2004 56, 423-52.

[37] Hedeker D; Mermelstein R J; Demirtas H. An application of a mixed-effects scale model for analysis of ecological momentary assessment (EMA) data. *Biometrics,* 2008 64, 627-34.

[38] . Dunton GF; Kawabata K; Intille S; Wolch J; Pentz, M. Assessing the social and physical contexts of children's leisure-time physical activity: An Ecological Momentary Assessment study.American Journal of Health Promotion, in press.

[39] Levav J; Fitzsimons GJ. When questions change behavior: the role of ease of representation. *Psychological Science,* 2006 17,207-13.

[40] Rowan PJ; Cofta-Woerpel L, Mazas CA; Vidrine JI; Reitzel LR; Cinciripini PM; Wetter DW. Evaluating reactivity to ecological momentary assessment during smoking cessation. *Experimental and Clinical Psychopharmacology.* 2007 15,382-9.

[41] Heron KE; Smyth JM. Ecological momentary interventions: incorporating mobile technology *British Journal of Health Psychology,* 2010 15, 1-39.

[42] Hurling R; Catt M; Boni MD; Fairley BW; Hurst T; Murray P; Richardson A; Sodhi JS. Using internet and mobile phone *Journal of Medical Internet Research,* 2007 9,7.

[43] Patrick K; Raab F; Adams MA; Dillon L; Zabinski M; Rock CL; Griswold WG; Norman GJ. A text message-based intervention for weight loss: randomized controlled trial. *Journal of Medical Internet Research,* 2009 11, 1.

In: Childhood Obesity: Risk Factors, Health Effects...   ISBN: 978-1-61761-982-3
Editor: Carol M. Segel   ©2011 Nova Science Publishers, Inc.

Chapter VIII

Effects of Schoolyard's Attributes on Childhood Obesity

Isabel Mourão-Carvalhal and Eduarda Coelho
University of Trás-os-Montes e Alto Douro, CIDESD

Abstract

Childhood obesity is a multifactorial disease resulting from an imbalanced energy intake and expenditure. The results from some investigations demonstrated that children are more likely to be active during school recess time; for that reason school playgrounds offer a sustainable context to increase vigorous physical activity in a way to counterbalance children's inactivity and to decrease obesity. The purpose of this study is to explore the association among obesity and gender, school year, physical activity, time spent going to school and how, areas and equipment of the schoolyard.

The sample included 975 children (505 girls and 470 boys) (age 8,28 + 1,23) from elementary school. Obesity was estimated by BMI and the cut off points of Cole et al. (2000). A questionnaire was completed by parents to provide information about age, school year, gender, time spent going to school and how, as well as the practise of physical activity. The areas (total and children per m2) from the schoolyards were calculated using the AutoCAD software. The schoolyards were characterized according to the equipments into four categories: non equipment; playground equipment; sport equipment and playground+sport equipment. The logistic regression was used to estimate the magnitude of association among overweight plus obesity and independent variables.

The results from multinominal logistic regression were only significant for time spent going to school (OR=0,612; 95% CI 0,437 – 0,856), total area (OR=1,756; 95% CI 1,036 – 2,975) and type of equipment - non equipment (OR=1,723; 95% CI 1,084 – 2,740) and playground equipment (OR= 2,048; 95% CI 1,063-3,947). The odds ratio of being obese was 0,612 times less for children that spend less than 10 minutes going to school. On the other hand, schoolyard area is a risk factor for obesity, 1,756 times more for children that play in a schoolyard with an area less than 1000 square meters. Also the absence of physical structures and the type of equipment are risk factors. The schoolyards that don't have any kind of equipment have a risk factor of 1,723 more to have obese children and those with playground equipment have a 2,048 higher risk than schools with a variety of equipment.

The results from this study highlighted the importance of schoolyards designed in a way so that the increase of physical activity and to reduce obesity. To promote healthy lifestyles, we suggest a multidisciplinary team to design playgrounds, taking into account the area and the physical structures.

Introduction

Childhood obesity is nowadays considered one of the major public health problems worldwide and more industrialized countries present higher prevalence than non developed countries. Between 1970 and 1990 the prevalence of overweight and obesity in school-age children has doubled or tripled in most parts of the world. The prevalence of overweight and obesity is not uniform; Europe and parts of the western Pacific and North America have the highest prevalence of overweight in the order of 20-30% (Wang & Lobstein, 2006). The lowest prevalence is noticed in the areas of southwestern Asia and Sub-Saharan Africa (Wang & Lobstein, 2006). According to the authors, by holding the same secular trend, it is estimated that by 2010 one in seven American children will be obese, according to the IOTF criteria as well as, one in ten children from European and Eastern Mediterranean regions present obesity. A look upon Europe's lower prevalence in countries from central and eastern Europe, and higher in southern countries are recorded, in the order of 20% to 40% in countries around the Mediterranean (Frelut & Lobstein, 2003). Portugal, according to the first national study, has a prevalence rate of 31.6% (Padez et al., 2004) occupying the second place in Europe, compared with other European countries: Italy (36%), Greece (31%) and Spain (30%), (www.ioft.org / childhood.euapendix.htm).

The major health consequences linked to a high prevalence of pediatric obesity are associated with psychological, cardiovascular risk, asthma, chronic inflammation, diabetes, orthopedic problems and liver disease (Reilly et al., 2003).

Childhood obesity is a multifactorial disease resulting from an imbalance between energy intake and energy expenditure. Causes associated with this imbalance are related to the interaction between genetic and environmental factors, which can be grouped into: biological, behavioral, social and environmental (Katzmarzyk et al., 2008).

Social, ecological and environmental approaches of health promotion are recommended in the study of complex interactions affecting individual lifestyles behaviors. According to social-ccological models, an individual's eating and physical activity behavior must reflect complex causes, considering the interactions between the inner/social personal layers and the influences of moderating environmental contexts. Bronfenbrenner's bio-ecological model (1994) conceptualizes the ecological environment as a set of nested structures at four successively more encompassing levels ranging from the micro to the macro. The microsystem is a proximal environment within which the child interacts with parents, siblings, peers and teachers. Examples include settings like home, workplace, neighborhood and school (Bronfenbrenner, 1994).

More recent investigations are now focused on environmental factors into the study of obesity, special physical and built environment (Trasande et al., 2009).

The built environment has been under investigation, recognizing that it plays an important role in increasing energy intake and reducing energy expenditure (Hill et al., 2003). In a meta-analysis Pappas et al. (2007), found that 84% of the articles emphasized a positive statistical association between some aspect of built involvement and obesity. Like teenagers and adults, children may be more influenced by involvement that is more immediate than what is further away (Pappas et al., 2007).

Neighborhood, school and home are three settings that can have an influence on physical activity behavior. The school has been identified as a key arena in promoting more active lifestyles. The students are the targets and spend large parts of their day at school; the teachers are the trainers, curriculum offers the training content and the built environment for the intervention. School playgrounds offer a sustainable context to increase vigorous physical activity in a way that counterbalances children's inactivity and decreases obesity. Schoolyards interventions could be important to increase recess physical activity and energy expenditure. The results from some investigations demonstrated that children are more likely to be active during school recess time, when there is a large number of activity-related equipment available, (Verstraete et al., 2006), (balls, hoops), permanent play (swings, climbing racks) and sports equipment (basquetball hoops) (Sallis et al., 2001). Data from intervention examining the effect of playground painting, such as hopscotch and small pieces for sports equipment, increases moderate to vigorous physical activity (Stratton et al., 2005). Children have adequate space to move freely and develop fundamental motor skills through play-based learning activities (NASPE, 2004).

Learning and practicing physical skills requires open space, such as a large room, gym, spacious hallway, or outdoor area.To incorporate active motor play, 100 square feet (30,48m2) per child is a reasonable minimum area for playgrounds (Greenman, 1988). Children must be able to swing a bat, kick a ball, or jump rope in a safe environment.

Playgrounds are less structured places that provide conditions for the practice of moderate physical activity through free play and games (Pate et al., 1996). Playgrounds are privileged places, where children spend much of their free time playing on spontaneous and exploratory games (Neto, 1997) and also in more vigorous motor activities (Ridgers et al., 2006). However, in general, the quality of these places and equipment is poor (Neto, 1997), with a small area, unattractive and low level of equipment diversity, enhancing anti-social behaviors associated with bullying (Pereira et al., 2002).

The various interventions during recess to improve the type of equipment (fixed and / or mobile) show an increased level of physical activity (Ridgers et al., 2007, Verstraete et al., 2006; Stratton & Mullan, 2005). Children perceive the function of new equipment, exploit and use it in their games. In contrast, the results of a study by Cardon et al. (2008) demonstrated an absence relationship between higher levels of physical activity and the existence of equipment such as slides and swings.

The purpose of this study is to explore the association among obesity and gender, school year, physical activity, time spent going to school and how, areas and equipment of schoolyard. Specifically to determine the risk and preventive factors of childhood obesity related to the schoolyard's attributes.

Materials and Methods

Sample

The sample included 858 children (441 girls and 417 boys) from elementary school. The mean age was 8,28 (±1,24), range between 5,40-12,00.

Measures and Procedures

In each school two trained persons measured height and weight using a standardized procedure. Height was measure with a stadiometer, with the lead in the Frankfort horizontal plane; weight was measure with an electronic scale with a precision of 100g. The BMI was calculated, weight/height 2 (kg/m2). Cut-offs were used to classify children as normal weight, overweight or obese (Cole et al., 2000). The cut-off points are linked to the widely accepted adult criteria for overweight (25kg/m2) and obesity (30kg/m2) and recent studies have shown high correlations between BMI and percent body fat measured by dual energy X-ray absorptiometry (DEXA).

Parents completed a questionnaire to provide information about age, school year, gender, by what means and how much time is spent going to school, and the practice of physical activity.

The total areas from the schoolyards were measured using AutoCAD software. The area per children was calculated dividing the total area by the total number of school children. The schoolyards were characterized according to equipment: no equipment, playground equipment (swings, slides and wheelies), sport equipment and playground+sport equipment.

The multinominal logistic regression was used to estimate the magnitude of association among overweight+obesity and independent variables.

Results

Descriptive Data

From the total 858 children included in the sample, 277 (32,3%) were classified as overweight and obese. The BMI mean was 17,86 (±2,94), with range between 12,00 and 30,6 (kg/m^2).

Relationship between Obesity and Schoolyard Variables

Table 1 illustrates the impact of the individual and schoolyards' variables in the prevalence of overweight and obesity in schoolchildren.

Table 1. Odds ratios and their respective 95% confidence interval of the prevalence of overweight and obesity for schoolchildren according to individual and schoolyards' variables

Variables		N	Overweight+ Obesity (%)	Odds ratios (95% CI)	p
Gender	boys	417	31,7	0,989 (0,738-1,326)	0,943
	girls	441	33,3	1	
Grade	1st	171	34,7	1,005 (0,656-1,541)	0,981
	2st	240	29,8	0,705 (0,471-1,055)	0,090
	3th	222	29,9	0,814 (0,542-1,222)	0,320
	4th	225	36,3	1	
Sport	Yes	337	30,2	0,861 (0,630-1,177)	0,349
	No	521	33,8	1	
Time spent going to school	≤ 10 minutes	626	30,3	0,612 (0,437-0,856)	0,004
	> 10 minutes	232	38,9	1	
Transportation go to school	toon foot	266	31,4	0,921 (0,656-1,292)	0,634
	motor vehicles	592	33,2	1	
Total schoolyards' area	< 1000 m^2	587	34,6	1,756 (1,036-2,975)	0,036
	≥ 1000 m^2	271	27,3	1	
Area per children	< 7 m^2	290	32,0	1,227 (0,647-2,327)	0,532
	7 - 17 m^2	263	33,6	1,063 (0,651-1,737)	0,807
	17 - 22 m^2	58	36,1	0,959 (0,445-2,065)	0,915
	> 22m^2	247	31,6	1	
Type of equipment	no equipment	241	35,1	1,723 (1,084-2,740)	0,021
	playground equipment	278	32,8	2,048 (1,063-3,947)	0,032
	sport equipment	179	37,4	1,916 (0,772-4,756)	0,161
	playground+sport equipment	160	25,4	1	

The results were only significant for: time spent going to school (OR=0,612; 95%CI 0,437–0,856), total area (OR=1,756; 95%CI 1,036–2,975) and type of equipment- no equipment (OR=1,723; 95%CI 1,084–2,740) and playground equipment (OR= 2,048; 95%CI 1,063-3,947).

The prevalence of overweight and obesity was higher in children that spend more than 10 minutes going to school than the others that spend less than 10 minutes (38,9% vrs. 30,3%). The odds ratio of being obese were 0,612 times less for children that spend less than 10 minutes going to school.

The prevalence of overweight and obesity was higher in children that play in schoolyards with an area less than 1000 m2 than those who play in an area more than 1000 m2 (34,6% vrs. 27,3%). Schoolyard area is a risk factor for obesity, 1,756 times more for children that play in a schoolyard with an area lower than 1000 m2.

The prevalence of overweight and obesity was higher in children that play in schoolyards with no equipments (35,1%) and playground equipment (32,8%) than schoolyards with more equipments (25,4%). The odds ratio of being obese was 1,723 times more for children that play in a schoolyard with no equipment and 2,048 times more for children that play in schoolyard with playground equipment.

Discussion

Regarding the prevalence of overweight and obesity, we can see that it is high (32,3%) and it has increased compared with previous data relating to the first national Portuguese study, for the same age group (31.6%) (Padez et al., 2004). When we consider the predictions for 2010 (Frelut & Lobstein, 2003), we find that this prevalence is within the limits expected by the authors, 20% to 40% for the Mediterranean and Southern European countries, where Portugal is located. There seems to be no logical explanation for the higher obesity prevalence in southern European countries compared with the northern ones, taking into account the Mediterranean dietary pattern and the temperate climate, of these countries. However, significant changes have occurred in the Mediterranean healthy diet, consisting of fruit, vegetables and fish, and nowadays replaced by high-fat food, less healthy. Also, sedentary behaviors, characteristic of these countries, may be explained by a reduction in thermogenesis (reducing the need to expend energy to stay warm), and discomfort experienced in performing physical activities in hot climates. Another reason may be due to smaller stature, which features the children of southern Europe, contributing to reverse the increase of the magnitude of body mass index (Frelut & Lobstein, 2003).

The prevalence of overweight and obesity was higher in children that spend more than 10 minutes going to school than the others that spend less time (38,9% vrs. 30,3%). Today, most children go to school by car, and not on foot, because of security problems and the greater distance between school and home. The increasing use of the car results in a more sedentary lifestyle, associated with a higher prevalence of obesity (Lopez, 2004, Frank et al., 2004). For each additional daily hour in the car it causes an increase of 6% of obesity, as well as, each mile a day walking is associated with a reduction of 4.8% of obesity in adults (Frank et al., 2004). The urban sprawl has led to a less independent mobility in children, also involving a reduction in games and physical activity outdoors (Veitch et al., 2007; Neto, 1997, 1999). The affordances perceived by children lead to a more sedentary behavior, such as watching television (Roemmich et al., 2007), Web surfing and playing computer games, unlike time outdoors, related to more vigorous activities (Hume et al., 2005).

The time spent in travelling to school is a risk factor for being obese. The children who spend more than 10 minutes 0,612 more chance to be obese.

The prevalence of overweight and obesity was higher in children that play in schoolyards with an area less than 1000 m2 (34,6% vs. 27,3%). The school has been identified as a key arena for promoting more active lifestyles. The results of Cardon et al. (2008), with pre-school children, showed a positive association between playground areas per child and the rate of physical activity, assessed by pedometers. The playgrounds are privileged places for the promotion of moderate physical activity through free play, in less structured environments (Pate et al., 1996). More space for children to play causes more energy expenditure and a better balance between energy intake and expenditure, resulting in a lower obesity. The results from this study show that schoolyard area is a risk factor for obesity, 1,756 times more for children that play in a schoolyard with an area lower than 1000 m2. According to Greenman (1988) a reasonable minimum area for playgrounds to promote play, is 100 square feet (30,48m2) per child.

Besides the total area, our results also demonstrate that the type of schoolyards' equipment is a risk factor in obesity prevalence. The lowest prevalence of obesity (25.4%)

was observed in schools with more equipment, so most important is not the quantity and type of equipment, but the diversity. As in the present study, the results of Mourão-Carvalhal et al. (2009) also show that a lower prevalence of obesity was found in schoolyards with more varied equipment. The quality of the places may be determinant in the level and diversity of play (Fjortoft, 2001, 2004), most creative games (Fjortoft, 2001) and higher rates of physical activity (Boldemann et al., 2005).

The risk of being obese is higher in children from schoolyards with playground, than those from schools with no equipment (2,048 vs. 1,723). The biggest risk associated with playground can be justified because swings, slides and wheelies are equipments which provoke dizziness and involve low energy expenditure, in opposition to an open place, without any equipment, that promotes games of rough and play, characteristics of these ages. Cardon et al. (2008) confirmed no relationship between higher levels of physical activity and the availability of equipment such as slides and swings.

As was mentioned previously, obesity should be conceptualized according to a social and ecological model, not being restricted to a micro level, such as school. We are conscious that other contexts should be analyzed (meso, exo and macrosystem), as well as the interactions between them. We suggest that other settings of the micro-level such as the house, neighborhood and peers should be included in the analysis of the obesity issue.

Conclusion

The results from this study highlight the importance of schoolyards' design so that physical activity is increased and obesity is reduced. Reduced areas and lower variability of equipments are risk factors, increasing the odds of children being obese. To promote healthy lifestyles, we suggest a multidisciplinary team to design playgrounds.

References

Boldemann, C., Blennow, M., Dal, H., Mårtensson, F., Raustorp, A., Yuen, K., & Wester, U. (2005). Impact of preschool environment upon children's physical activity and sun exposure. *Preventive Medicine, 42*, 301-308.

Bronfenbrenner, U. (1994). Ecological models of human development. *In International Encyclopedia of Education,* Vol.3, 2nd. Ed. Oxford: Elsevier. Reprinted in: Gauvain, M. & Cole, M. (Eds.), readings on the development of children, 2nd Ed. (1993, pp.37-43). NY: Freeman.

Cardon, G., Cauwenberghe, E., Labarque, V., Haerens, L. & Bourdeaudhuij, I. (2008). The contribution of preschool playground factors in explaining children's physical activity during recess. *Int J Behav Nutr Phys Act, 26*, 5–11.

Cole, T., Bellizzi, M., Flegal, K. & Dietz, W. (2000). Establishing a standard definition for child overweight and obesity worldwide: international survey. *British Medical Journal, 320*, 1240-1243.

Fjortoft, I. (2001). The natural environment as a playground for children: the impact of outdoor play activities in pre-primary school children. *Early Childhood Education Journal, 29,* 111-117.

Fjortoft, I. (2004). Landscape as playscape: the effects of natural environments on children's play and motor development. *Children, Youth and Environments, 14,* 21-44.

Frank, L., Anderson, M., Thomas, MA, & Schmidt, L. (2004). Obesity relationships with community design, physical activity, and time spent in cars. *Am J. Prev Med, 27,* 87-96.

Frelut, M. & Lobstein, T (2003). Prevalence of overweight among children in Europe. *Obesity Reviews, 4,* 195-200.

Greenman, JT (1988). Caring spaces, learning places. Redmond, WA: Exchange Press.

Hill JO, Wyatt HR, Reed GW et al. (2003). Obesity and the environment: where do we go from here? *Science, 299,* 853-5.

Hume, C., Salmon, J. & Ball, K. (2005). Children's perceptions of their home and neighborhood environments, and their association with objectively measured physical activity: a qualitative and quantitative study. *Health Education Resarch, Theory & Practice, 20,* 1-13.

Katzmarzyk, P., Baur, L., Blair, S., Lambert, E., Oppert, J. & Riddoch, C., (2008). International conference on physical activity and obesity in children: summary statement and recommendations. *International Journal of Pediatric Obesity, 3,* 3-21.

Lopez, R. (2004).Urban sprawl and risk for being overweight or obese. *American Journal of Public Health, 94,* 1574-79.

Mourão-Carvalhal, I, Coelho, E., Laranjeira, H., Monteiro, G. & Azevedo, A. (2009). Recreios escolares; Comparação das áreas e tipos de equipamentos em função da obesidade. In L. Rodrigues, L. Saraiva, J. Barreiros, & O. Vasconcelos, O. (Eds). *Estudos em Desenvolvimento Motor da Criança II* (pp. 217-224). Instituto Politécnico de Viana do Castelo.

NASPE. (2004). *Physical Activity for Children: A Statement of Guidelines for Children Ages 5-12* (2nd ed.). Reston, VA: NASPE Publications.

Neto, C. (1997). Jogo e desenvolvimento da criança - introdução. In C. Neto (Ed.). *O Jogo e o Desenvolvimento da Criança* (pp. 5-9). Lisboa: Edições FMH.

Neto, C. (1999). O jogo e os quotidianos de vida da criança. In R. Krebs, F. Copetti & T. Beltram (Eds.). *Perspectivas para o Desenvolvimento Infantil* (pp. 49-66). Santa Maria - Brasil: Edições SIEC - Santa Maria.

Padez, C., Fernandes, T., Marques, V., Mourão-Carvalhal, I. & Moreira, P. (2004). Prevalence of overweight and obesity in 7-9 year-old Portuguese children. Trends in body mass index from 1970-2002. *American Journal of Human Biology, 16,* 1-9.

Pappas, M., Alberg, A., Ewing, R., Helzisouer, K., Gary, T. & Klassen, A. (2007). The built environment. *Epidemiologic Reviews, 29,* 129-143.

Pate, R., Baranowski, T., Dowda, M. & Trost, S. (1996). Tracking of physical activity in young children. *Medicine Science Sports* Exercise, *28,* 92-96.

Pereira, B., Neto, C., Smith, P. K., & Angulo, J. (2002). Reinventar los espacios de recreo. Prevenir los comportamientos agresivos. *Cultura y Educación, 14-23,* 297-311.

Reilly JJ, Methven E, McDowell SC, et al. (2003). Health consequences of obesity: systematic review. *Arch Dis Child, 88,* 748-52.

Ridgers ND, Stratton G, Fairclough SJ (2006). Physical activity levels of children during school playtime. *Sports Med, 36,* 359-371.

Ridgers ND, Stratton G, Fairclough SJ, Twisk, J. (2007). Long-term effects of a playground markings and physical structures on children´s recess physical activity levels. *Prev Med*, *44*, 393-397.

Roemmich J, Epstein L, Raja S, Yin L, (2007). The Neighborhood and Home Environments: Disparate Relationships With Physical Activity and Sedentary Behaviors in Youth. *Ann Behav Med, 33(1)*, 29-38.

Sallis, J., Conway, D., Prochaska, J., Mckenzie, T. & Nelson, J. (2001). The association of school environments with youth physical activity. *American Journal of Public Health, 91*, 618-620.

Stratton G, Mullan E (2005). The effect of multicolor playground markings on children's physical activity level during recess. *Prev Med, 4,* 828-833.

Trasande, L., Cronk, C., Durkin, M., Weiss, M., et al (2009). Environment and obesiy in the national children's study. *Environmental Health Perspective, 117*, 159-166.

Veitch, J., Salmon, J. & Ball, K. (2008). Cildren's active free play in local neighborhoods: a behavioral mapping study. *Health Education Research, 23*, 870-879.

Verstraete, S.J., Cardon, G.M., De Clercq, D.L.R., & De Bourdeaudhuij, I.M.M. (2006). Increasing children's physical activity levels during recess periods in elementary schools: the effects of providing game equipment. *European Journal of Public Health, 16,* 415–419.

Wang, W. & Lobstein, T. (2006). Worldwide trends in childhood overweight and obesity. *International Journal of Pediatric Obesity, 1*, 11-25.

In: Childhood Obesity: Risk Factors, Health Effects...   ISBN 978-1-61761-982-3
Editor: Carol M. Segel   © 2011 Nova Science Publishers, Inc.

Chapter IX

A GENDER-STRUCTURED RANDOM MODEL FOR UNDERSTANDING ADULTHOOD OBESITY

Francisco-J. Santonja[1], *Rafael-J. Villanueva*[2]* *and Lucas Jódar*[2]
[1] Departamento de Estadística e Investigación Operativa
Universidad de Valencia, Dr. Moliner 50, 46100 Burjassot (Valencia), Spain
[2] Instituto de Matemática Multidisciplinar, Edificio 8G piso 2
Universidad Politécnica de Valencia, 46022 Valencia, Spain

Abstract

Excess weight population is increasing in developed and developing countries and are becoming a serious health concern. The Spanish Society for the Study of Obesity (SEEDO) indicates the importance of addressing it on a high-priority basis. In this article, we propose a mathematical model of epidemiological type to predict the incidence of excess weight in 24- to 65-year-old females and males in the region of Valencia (Spain) in the coming years. Due to the intrinsic uncertainty of the statistical data used, we also apply the Latin Hypercube Sampling (LHS) to the model in order to obtain predictions over the next few years providing 90% confidence intervals.

1 Introduction

The obesity epidemic is a present and potential public health concern. In the region of Valencia (Spain) it is becoming a serious problem not only from an individual health point of view but also from the public socioeconomic one. Note that valencian obese population is increasing [1, 2] (see Table 1).

In this paper, we develop an epidemic mathematical model (based on differential equations) to study the evolution of the obesity epidemic in the region of Valencia (Spain) over the next few years. As it has been proposed in [3, 4] we consider obesity as a disease of social transmission. The main idea behind this approach is that unhealthy nutritional habits and sedentary lifestyles may spread from one person to another.

---

*E-mail addresses: francisco.santonja@uv.es, rjvillan@imm.upv.es, ljodar@imm.upv.es

Table 1: Population in the region of Valencia in 2000 and 2005 divided by their Body Mass Index (BMI): Normal weight for BMI<25, Overweight for 25≤BMI<30 and Obese for BMI≥30 [1, 2].

2000	Normal weight	Overweight	Obese
Feminine Population	35.0%	17.8%	7.2%
Masculine Population	17.1%	18.4%	4.5%
Total	52.1%	36.2%	11.7%

2005	Normal weight	Overweight	Obese
Feminine Population	31.0%	14.0%	7.5%
Masculine Population	17.0%	24.0%	6.5%
Total	48.0%	38.0%	14.0%

Systems of ordinary differential equations have become important tools in analysing the spread and control of infectious diseases. Special models have been formulated for diseases such as rubella, HIV, malaria or syphilis [5] although recently several differential equation models have also been proposed for modeling social behavior such as ideologies or drug abuse [6, 7, 8].

2   Mathematical model

To build this epidemiological model, Valencian individuals (24-65 years old) are divided into three subpopulations using their Body Mass Index ($BMI = weight/height^2$): N_j, individuals with normal weight; S_j, individuals with overweight and O_j, obese individuals (j=1: Feminine population; j=2: Masculine population).

We build this epidemiological mathematical model considering the following assumptions:

1. We assume population homogeneous mixing. That is, each individual can contact with any other individual of any other subpopulation [9].

2. We consider that sedentary lifestyles and unhealthy nutritional habits increase the individual weight of the adults in the region of Valencia [10]. Hence, the progression to overweight subpopulation (S_j) or obese subpopulation (O_j) are determined by the transmission of unhealthy lifestyles (sedentarism and unhealthy nutritional habits) from one person to another.

3. The transitions described can be modelled as follows:

   - An individual in N_j transits to S_j because people in S_1, S_2, O_1 or O_2 transmit unhealthy lifestyles by social contact at rate β_j, $j = 1,2$. This transit is modelled using the term $\beta_j N_j (S_1+S_2+O_1+O_2)$.

- An individual in S_j transits to O_j at rate γ_j if he/she maintains his/her unhealthy lifestyle ($\gamma_j S_j$).
- An individual in S_j or O_j transits to N_j or S_j respectively, when he/she decides to give up his/her unhealthy lifestyle and starts with the practice of physical activity and puts on diet. These transits are modelled by $\rho_j S_j$ and $\varepsilon_j O_j$.

4. Individuals leave the system when they are 65 years old proportionally to the size of the subpopulations N_j, S_j and O_j, and modelled by μN_j, μS_j and μO_j, where μ is the time an individual stays in the system, i.e., 42 years.

5. We consider that the newly recruited 24 years old individuals become members of N_j, S_j or O_j proportionally to their sizes, i.e., modelled by the terms μN_j, μS_j and μO_j.

6. We also assume that mortality rates can be disregarded.

These three last premises imply that women and men subpopulations are both constant over the time.

Under the above assumptions, the excess weight transmission dynamics is described by the following system of differential equations (t, time in weeks).

$$
\begin{aligned}
N_1'(t) &= -\beta_1 N_1(t)\Big(S_1(t)+O_1(t)+S_2(t)+O_2(t)\Big)+\rho_1 S_1(t)\\
&\quad +\mu N_1(t)-\mu N_1(t)\\
S_1'(t) &= \beta_1 N_1(t)\Big(S_1(t)+O_1(t)+S_2(t)+O_2(t)\Big)-\rho_1 S_1(t)\\
&\quad +\varepsilon_1 O_1(t)-\gamma_1 S_1(t)+\mu S_1(t)-\mu S_1(t)\\
O_1'(t) &= \gamma_1 S_1(t)-\varepsilon_1 O_1(t)+\mu O_1(t)-\mu O_1(t)\\
N_2'(t) &= -\beta_2 N_2(t)\Big(S_1(t)+O_1(t)+S_2(t)+O_2(t)\Big)+\rho_2 S_2(t)\\
&\quad +\mu N_2(t)-\mu N_2(t)\\
S_2'(t) &= \beta_2 N_2(t)\Big(S_1(t)+O_1(t)+S_2(t)+O_2(t)\Big)-\rho_2 S_2(t)\\
&\quad +\varepsilon_2 O_2(t)-\gamma_2 S_2(t)+\mu S_2(t)-\mu S_2(t)\\
O_2'(t) &= \gamma_2 S_2(t)-\varepsilon_2 O_2(t)+\mu O_2(t)-\mu O_2(t)
\end{aligned}
$$

The model diagram is depicted in Figure 1. The boxes represent the subpopulations and the arrows represent the transitions between the subpopulations. Arrows are labelled by their corresponding model parameters.

The parameters of the model are:

- μ, average stay time in the system of 24-65 years old adults.

- β_j, $j=1,2$, transmission rate to adopt unhealthy lifestyles (unhealthy nutritional habits and sedentary lifestyles).

- γ_j, $j=1,2$, rate at which an overweight 24-65 years old adult becomes an obese individual because of unhealthy lifestyles.

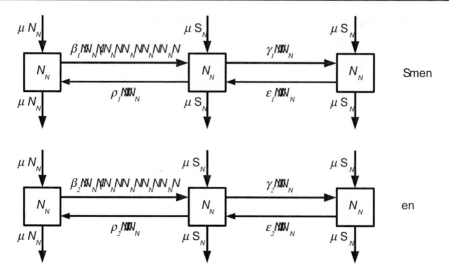

Figure 1: Flow diagram of the gender-structured mathematical model for the dynamics of excess weight prevalence in the population of the region of Valencia.

- $\rho_j = \rho_j^1 * \rho_j^2 * p$, $j=1,2$, rate at which an overweight individual (putting on diet and practising physical activity) moves to normal weight individual.

  - ρ_j^1 measures the effectiveness of physical activity in overweight population to reduce weight.
  - ρ_j^2 measures the effectiveness of diet in overweight population to reduce weight.
  - p measures the time an individual needs to return to overweight subpopulation (from obese subpopulation) or normal weight subpopulation (from overweight subpopulation) by diet and physical activity. We assume that this parameter is independent of gender.

- $\varepsilon_j = \varepsilon_j^1 * \varepsilon_j^2 * p$, $j=1,2$, rate at which an obese individual (putting on diet and practising physical activity) moves to overweight population.

  - ε_j^1 measures the effectiveness of physical activity in obese population to reduce weight.
  - ε_j^2 measures the effectiveness of diet in obese population to reduce weight.
  - p has been already described above.

The solution of this system of differential equations allows us to predict the proportion of overweight individuals or obese individuals (24-65 years old) in the next few years for each gender class.

3 Parameters estimation

We can obtain the parameters of the model except β_j and γ_j, $j = 1,2$, using the following sources:

- Health Survey for region of Valencia, year 2000 [1].

- Health Survey for region of Valencia, year 2005 [2].

- Two technical reports published by J.J. Arricabalaga et al. [11] and J. Salas et al. [12], respectively. In these reports appear some details about strategies to reduce BMI.

The parameters of the model are estimated as follows:

- $\mu = 1/2184 = 0.00045788$ $weeks^{-1}$. We consider that the mean time spent by an individual in the system is 42 years, i.e., 2,184 weeks.

- $\rho_j = \rho_j^1 * \rho_j^2 * p$, $j = 1, 2$. These parameters are factorized in three factors, ρ_j^1, ρ_j^2 and p. ρ_j^1 is the percentage of overweight subpopulation who practice physical activity, for females and males, ρ_j^2 is the percentage of overweight subpopulation who control their nutritional habits, for feminine and masculine subpopulations, and p is the time an individual needs to return to normal weight subpopulation by diet and physical activity. Taking into account the Health Survey for the region of Valencia (year 2000), we estimate ρ_j^1 and ρ_j^2. The values are: $\rho_1^1 = 0.00712$, $\rho_1^2 = 0.03916$, $\rho_2^1 = 0.00736$, $\rho_2^2 = 0.04048$.

  To estimate p, we consider that overweight individuals need to decrease a mean of 7 Kg. to move to normal weight population [1, 2]. We also know that a 24-week diet plus exercise program produces a weight loss of about 1 Kg. per week [11]. Then we consider $p = 1/7$ $weeks^{-1}$. We assume that this parameter is independent of gender.

- $\varepsilon_j = \varepsilon_j^1 * \varepsilon_j^2 * p$, $j=1,2$. These parameters are also factorized in three factors, ε_j^1, ε_j^2 and p. ε_j^1 is the percentage of obese subpopulation who practice physical activity, for feminine and masculine subpopulations, ε_j^2 is the percentage of obese subpopulation who control their nutritional habits, for females and males, and p is the time an individual needs to return to overweight subpopulation by diet and physical activity. Taking into account the Health Survey for the region of Valencia (year 2000), we estimate ε_j^1 and ε_j^2. The values are: $\varepsilon_1^1 = 0.00288$, $\varepsilon_1^2 = 0.01584$, $\varepsilon_2^1 = 0.00176$, $\varepsilon_2^2 = 0.00968$. In this case, we also consider $p = 1/7$ $weeks^{-1}$.

- Taking into account as initial conditions N_j, S_j and O_j in 2000 and as final conditions N_j, S_j and O_j in 2005 (see Table 1) parameters β_j and γ_j have been estimated by fitting the model with data.

In order to compute the best fitting, we carried out computations with *Mathematica* [13] and we implemented the function:

$$\mathbb{F} : \quad \mathbb{R}^4 \quad \longrightarrow \quad \mathbb{R}$$
$$(\beta_j, \gamma_j) \quad \longrightarrow \quad \mathbb{F}(\beta_j, \gamma_j)$$

whose variables are β_j ($j = 1, 2$) and γ_j ($j = 1, 2$), such that:

1. Solve numerically (*NDSolve[]*) the system of differential equations with initial values in year 2000 (see Table 1),

2. For $t = 260$ (year 2005) evaluate the computed numerical solution for each subpopulation $N_j(t)$, $S_j(t)$ and $O_j(t)$.

3. Compute the mean square error between the values obtained in Step 2 and the data of year 2005 (see Table 1).

Function $\mathbb{F}$ takes values in $\mathbb{R}^4$ (β_j, γ_j, $j=1,2$) and returns positive real numbers. Hence, we can try to minimize this function using the Nelder-Mead algorithm, that does not need the computation of any derivative or gradient, impossible to know in this case [14, 15]. Then, the values of β_j and γ_j that minimize the function $\mathbb{F}$ are:

$$\beta_1 = 0.000312, \ \beta_2 = 0.001564, \ \gamma_1 = 0.000603, \ \gamma_2 = 0.000121.$$

These parameters, on one hand, indicate that the transmission rate in women is less than the one in men ($\beta_1 < \beta_2$), i.e., men are more susceptible to adopt unhealthy lifestyles than women. On the other hand, the transition from overweight to obese is more fluid in women ($\gamma_1 > \gamma_2$), but it is compensated in some way by the return parameter ε_j, greater for women ($j = 1$) than men ($j = 2$).

4   Trends over next few years

4.1   Deterministic prediction

Once the model parameters have been estimated, under the assumption that the public health policies related to change lifestyles with effects on the population weight do not vary, we can use the model to predict the trend of each subpopulation over the next few years. The model forecasting for years 2011, 2013 and 2015 can be seen in Table 2.

Table 2: Deterministic prediction for next years 2011, 2013 and 2015.

	N women	S women	O women	N men	S men	O men
2011	32.23%	14.99%	12.78%	11.11%	22.95%	5.94%
2013	31.69%	14.6%	13.7%	10.22%	23.56%	6.23%
2015	31.16%	14.25%	14.6%	9.38%	24.08%	6.53%

Looking at Tables 1 and 2 we may infer that:

- the model underestimates the excess weight population in both sexes because data in 2005 are greater than in some of the following years,

- men excess weight subpopulations are increasing,

- a worrying increasing can be seen in women obese subpopulation,

- women overweight subpopulation has a slight decreasing,

- men normal subpopulation is decreasing faster than women normal subpopulation,

- if the health policies do not change over the next few years, the model predicts 56.66% in 2011, 58.09% in 2013 and 59.46% in 2015 of people in the region of Valencia will be overweight or obese.

4.2   Random prediction

Although the analysis developed above gives us some interesting conclusion, the data used in the model parameter estimation contain uncertainties, and, therefore, the supposition that parameters always remain constant or data from Health Surveys do not contain errors, is not appropriate. Hence, the deterministic prediction can give us an idea about future trends but the obtained values may not be as accurate as we expect.

To compensate for this drawback, we propose forecasting future trends using confidence intervals. In order to calculate these confidence intervals, let us use the technique called Latin Hypercube Sampling (LHS) to vary parameter values in the proposed model. LHS, a type of stratified Monte Carlo sampling, is a sophisticated and efficient method for achieving equitable sampling of all input parameters simultaneously [16, 17]. Each parameter for a model can be defined as having an appropriate probability density function associated with it. It is usual to use the uniform distribution centred at deterministic parameter estimators in absence of data to give information as to the distribution for a given parameter [17, 18]. Thus, the model can be simulated by sampling a single value from each parameter distribution. Many samples should be taken and many simulations should be run, producing variable output values that can be treated with descriptive statistic techniques to compute the means and 90% confidence intervals.

Table 3 shows details about the variation ranges of the input parameters. The variation intervals for β_j and γ_j are defined assuming that the value of these parameters may have a rounding perturbation not greater that 20%. The variation range for the other parameters is defined by the binomial proportion confidence intervals [19].

Table 3: Parameter variations

Parameter	Interval	Parameter	Interval
ρ_1^1	$[0.004612, 0.009627]$	ρ_1^2	$[0.03337, 0.04494]$
ρ_2^1	$[0.00481, 0.0099]$	ρ_2^2	$[0.0346, 0.0463]$
ε_1^1	$[0.00128, 0.00447]$	ε_1^2	$[0.01211, 0.01956]$
ε_2^1	$[0.000509, 0.003019]$	ε_2^2	$[0.00675, 0.01260]$
β_1	$[0.0002496, 0.0003743]$	β_2	$[0.001251, 0.001876]$
γ_1	$[0.0004825, 0.0007238]$	γ_2	$[0.00009677, 0.0001451]$
p	$[1/9, 1/5]$		

Thus, the model can be simulated by sampling a single value from each input distribution. In our case, LHS is used to generate $5,000$ different values of each input value (transmission parameters and initial conditions). We used these samples to run 5 ,000 evaluations of the deterministic value. The results of these evaluations allow us to determinate the 90% confidence intervals output taking into account the empirical 5% and 95% percentiles. In the definition of the output 90% confidence intervals we consider the following binomial proportion confidence intervals for the initial conditions:

$$N_1^0 \in [0.3357, 0.3642],\ S_1^0 \in [0.1665, 0.1894],\ O_1^0 \in [0.0642, 0.0797],$$
$$N_2^0 \in [0.1597, 0.1822],\ S_2^0 \in [0.1724, 0.1955],\ O_2^0 \in [0.0388, 0.0511].$$

These intervals have been defined taking into account the Health Survey for the Region of Valencia (year 2000). The 90% confidence intervals from the 5,000 outputs obtained for the coming years 2011, 2013 and 2015 can be seen in Table 4.

Table 4: 90% confidence intervals obtained with 5,000 model outputs using LHS technique over the next years 2011, 2013 and 2015.

	N women	S women	O women
2011	[30,90%, 33.61%]	[13.69%, 16.32%]	[11.54%, 13.98%]
2013	[30.33%, 33.14%]	[13.21%, 16.02%]	[12.35%, 15.00%]
2015	[29.75%, 32.66%]	[12.78%, 15.75%]	[13.15%, 15.99%]

	N men	S men	O men
2011	[9.97%, 12.43%]	[21.32%, 24.52%]	[5.29%, 6.57%]
2013	[9.04%, 11.61%]	[21.81%, 25.19%]	[5.54%, 6.89%]
2015	[8.18%, 10.83%]	[22.26%, 25.78%]	[5.79%, 7.22%]

If there are not changes in current health policies in the next few years, the mathematical model presented predicts that around 59.36% ($90\% CI = [53.98\%, 64.74\%]$) of 24-65 years old inhabitants of the region of Valencia will present excess weight in 2015.

5   Conclusions

The mathematical model presented shows that the incidence of excess weight is increasing in the region of Valencia: 47.9% in 2000, 52.0% in 2005, 56.66% in 2011, 58.09% in 2013 and 59.46% in 2015. Additionally, the estimation of parameters β_1 and β_2 provides crucial information about the adoption of unhealthy lifestyles: masculine population is the most influenced sex ($\beta_2 > \beta_1$). Therefore, we may conclude that masculine population is a strategic population to focus preventive actions to address the obesity epidemic in the region of Valencia.

Other point that should be taken into account is the transition of women from overweight to obese subpopulation where treatment policies could reduce the worrying prediction over the next few years for women obese subpopulation.

In this modelling approach, we show how epidemiological mathematical models can be an useful tool to understand the obesity epidemic. Using this type of mathematical models policy makers can provide insight into this problem.

6   Acknowledgments

The authors acknowledge the collaboration of the Valencian Department of Health and the Valencian Economic Research Institute (IVIE). This work has been supported by 2010 Contract for Research Program of the Valencian Economic Research Institute (IVIE).

References

[1] Valencian Health Department, Health Survey database, 2000, Available at http://www.san.gva.es/val/prof/homeprof.html

[2] Valencian Health Department, Health Survey database, 2005, Available at http://www.san.gva.es/val/prof/homeprof.html

[3] N. A. Christakis, J. H. Fowler, *Connected: the surprising power of our social networks and how they shape our lives*, Brown and Company. Hachette Book Group, 2009.

[4] N. A. Christakis, J. H. Fowler, The spread of obesity in a large social network over 32 years, *The New England Journal of Medicine* 357 (2007). 370–379.

[5] H. Hethcote, The mathematics of infectious diseases, *SIAM Rev.* 42 (2000), Issue 4, 599–653. (2000).

[6] F. J. Santonja, A. C. Tarazona, R. J. Villanueva, A mathematical model of the pressure of an extreme ideology on a society, *Computers and Mathematics with Applications* 56 (2008), 836–846.

[7] F. J. Santonja, E. Sánchez, M. Rubio, J. L. Morera, Alcohol consumption in Spain and its economic cost: a mathematical modeling approach, *Mathematical and Computer Modelling*, 2010, doi:10.1016/j.mcm.2010.02.029.

[8] E. Sánchez, R. J. Villanueva, F. J. Santonja, M. Rubio, Predicting cocaine consumption in Spain. A mathematical modeling approach, Drugs: Education, *Prevention & Policy*, 2010, Accepted. To appear.

[9] J. D. Murray, Mathematical Biology, Springer, 2000.

[10] F. J. Santonja, R. J. Villanueva, L. Jódar, G. González, Mathematical modeling of social obesity epidemic in the region of Valencia, Spain, *Mathematical and Computer Modelling of Dynamical Systems* 16 (2010), 23–34.

[11] J. J. Arrizabalaga, L. Masmiquel, J. Vidal, A. Calaas, M. J. Díaz, P. P. Garca, S. Monereo, J. Moreiro, B. Moreno, W. Ricart, F. Cordido, Recomendaciones y algoritmo de tratamiento del sobrepeso y la obesidad en personas adultas [Recomendations for overweight and obesity treatment in adulthood], *Medicina Clínica* 122 (2004), 104–110.

[12] J. Salas, M. A. Rubio, M. Barbany, B. Moreno, Consenso 2007 para la evaluación del sobrepeso y la obesidad y el establecimiento de criterios de intervención terapéutica [2007 Report: Overweight and obesity evaluation and establishment of criteria for therapeutic intervention], *Revista Española de Obesidad* 128 (2007), 184–196.

[13] *Mathematica*, htpp://www.wolfram.com/products/mathematica.

[14] J. A. Nelder, R. Mead, A simplex method for function minimization, *Computer Journal* 7 (1964), 308–313.

[15] W. H. Press, B. P. Flannery, S. A. Teukolsky, W. Vetterling, *Numerical recipes: the art of scientific computating*, Cambridge University Press, 1986.

[16] S. M. Blower, H. Dowlatabadi, Sensitivity and Uncertainty Analysis of Complex Models of Disease Transmission: and HIV Model, as an Example, *International Statistical Review* 62 (2) (1994), 229–243.

[17] A. Olsson, G. Sandberg, O. Dahlblom, On Latin hypercube sampling for structural reliability analysis. *Structural Safety* 25 (1) (2003), 47–68, doi: 10.1016/S0167-4730(02)00039-5.

[18] M. McKay, M. Meyer, Critique of and Limitations on the Use of Expert Judgements in Accident Consequence Uncertainty Analysis, *Radiation Protection Dosimetry* 90 (3) (2000), 325–330.

[19] R. V. Hogg, J. W. McKean, A. T. Craig, *Introduction to Mathematical Statistics*, Prentice Hall, 2005.

In: Childhood Obesity: Risk Factors, Health Effects…   ISBN: 978-1-61761-982-3
Editor: Carol M. Segel   ©2011 Nova Science Publishers, Inc.

Chapter X

Does This Study Inform Policy? Examination of Leading Childhood Obesity Journals' Instructions to Authors Regarding Policy-Related Research and Implications

Sheila Fleischhacker[*]*, Kelly R. Evenson, Puneet Singh and Daniel A. Rodriguez*
University of North Carolina-Chapel Hill, NC, USA

Abstract

This study examines leading childhood obesity journals' instructions to authors regarding policy-related research and implications. A systematic approach using five data sources was used to determine 50 leading childhood obesity journals. The following information was obtained from each journal: aims and scope; instructions to authors; types of articles published; and, each type of articles' format and abstract structure. Fifteen of the 50 journals (30%) included policy in their aims and scope. Eighteen journals (36%) explicitly instructed authors on a type of article where policy-related information could be published. Only three journals (6%) explicitly indicated a potential policy outlet within their research article format and one journal provided a potential policy outlet within their abstract structure. Opportunity exists for the development of more explicit editorial policies and best practices on how researchers incorporate policy components in their study design and articulate policy implications in their publications.

---

[*] Sheila Fleischhacker, PhD, JD, NIH-UNC Interdisciplinary Obesity Post-Doc , University of North Carolina, Gillings School of Global Public Health, Center for Health Promotion & Disease Prevention, CB#7426, 1700 Martin Luther King Jr. Blvd, Chapel Hill, NC 27599, Phone: 919-843-9563, Email: sheilafly9@gmail.com

Introduction

Policy strategies are increasingly playing a role in efforts to reduce childhood obesity. Urgency is needed since almost one third of American children over the age of two are overweight or obese(1). In a recent "Perspective" article published in the *New England Journal of Medicine* data was presented on the role of pregnancy and infancy in the formation of childhood overweight implying that the First Lady's Let's Move campaign (www.letsmove.gov) should expand beyond school-aged children and school-centered approaches(2). This is one example of how a peer-reviewed journal article served as an outlet for discourse on policy initiatives.

The role of policy discourse in peer-reviewed journals and in research papers specifically has been debated(3-11). Common themes emerging from this historical debate and more recent discussions on effectively translating evidence into policy(12-18) express the importance of maintaining the objectivity of the evidence-base, bridging the gap between researchers and policymakers, and addressing deficiencies in researchers, policymakers, and other stakeholders' understanding of how to ensure the evidence on an issue makes meaningful contributions to the policy process. Specific to childhood obesity, researchers(19-21) have called for increased rigor in studies investigating the role of policy in improving healthy eating and active living, as well as improved application of environmental and policy research to change policy and practice. Indeed, an important macro-level component of an ecological approach to health promotion and disease prevention includes implementing policies with the capacity to enhance health, create supportive environments, and strengthen community action(22). The emphasis reflected in research on obesity and funding priorities, however, traditionally targets only individual behavior change(19).

The purpose of this study was to examine leading childhood obesity journals' instructions to authors regarding policy-related research and implications. We hypothesized that several journals have aims to influence public health policy, but few journals within their guidance to authors would explicitly provide instruction on if and how submissions could address policy strategies from the public, private, or non-for-profit sectors. Our ultimate goal is to advance evidence-based childhood obesity policy by providing insight into how researchers can more effectively incorporate policy components in their study design and articulate policy implications in research articles and other communication venues.

Methods

Determining the Leading Childhood Obesity Journals

With the assistance of two Health Sciences Librarians, a systematic approach using five data sources was used to determine 50 leading childhood obesity journals.

1) PubMed

In November 2009, two researchers independently searched PubMed using the Medical Subject Heading (i.e., MeSH term) obesity. The search was restricted to include articles where obesity was a major aspect and the obesity term was not exploded. The search was limited to humans, English language, all child, and with publication dates between 1998/01/01 to 2009/11/20. This search generated 8,597 articles. To focus on obesity prevention and control, these articles were then filtered using the following exclusion: NOT (surgery OR surgical OR bariatric), which trimmed the list to 8,187 articles. Next, the resulting 8,187 articles were sorted by journal in alphabetical order. For each journal, one researcher tallied how many articles were listed and the second researcher checked this count using the PubMed search by journal option. These journal totals were used by each researcher to generate a list of journals with at least 30 articles. Independently, each researcher generated the same list of journals (n=40) with over 30 articles published.

2) Web of Science

For our second approach, two investigators independently obtained a list of journals frequently publishing childhood obesity articles within the Web of Science database in December 2009. The Web of Science search protocol used a modified PubMed search process, since this database does not use MeSH terms. Instead, (child*AND obesity) NOT (surg* or bariatric) was used. We limited our search to articles, excluding abstracts, and searched within the same publication dates (1998/01/01 to 2009/11/20). Unlike PubMed, the Web of Science generates a list of journals by article count. The Web of Science' similar findings provided further support of our list of 40 journals and only two more journals were added to our list.

3) Impact Factor

We used a journal's impact factor to understand the relative importance of each journal within its field. An impact factor is a measure reflecting the average number of citations to articles published in science and social science journals. The higher the journal's impact factor, the greater the importance of the journal. For each journal listed on our combined PubMed and Web of Science list, two researchers independently searched the Web of Science database for the 2008 impact factor. The impact factor of four journals was not found on Web of Science's database; therefore, the impact factor provided on each of these four journal's websites was used.

4) Best Evidence for Clinical Pediatrics Journal Rankings

Our fourth approach was to check our current list of 42 journals with the list of ranked journals in an article detailing which journals pediatricians will find the best evidence for clinical practice(23). One journal was added.

5) NCI Food Environment Database

Finally, two researchers worked together to determine the journal for each article listed (n=153) in the National Cancer Institute's (NCI) Food Environment database (https://riskfactor.cancer.gov/mfe/). For journals currently in our list, we noted the number of articles listed in NCI. We added three journals to our list, since relative to other journals found through the NCI database they had a large number of articles archived in the database.

After evaluating the collective information provided from these five approaches, our list included 46 journals. To finalize a list of 50 journals, we selected the 41st through 44th journals listed in our step one, PubMed article frequency list. We similarly used our other four approaches to evaluate these four additional journals. All journals examined were categorized (e.g., epidemiology or public health) using the journal's title and online information provided on the journal's website. In some instances, a journal was categorized into a primary and secondary category.

Examining the Policy Aims and Instructions to Authors

During January and February, 2010, two independent researchers obtained the following information from each journal's website: aims and scope; instructions to authors; types of articles published; and, each type of articles' format and abstract structure. All of the information on these topics provided by each journal was read and coded. Each researcher coded whether or not (yes or no) public health policy or related terms (e.g., law or regulation) were used to describe a journal's aims and scope (i.e., vision, mission, purpose, or objectives). If yes, the text was copied verbatim and the section where this text was located was noted. For our purposes, we considered public health policy to include any reference to or guidance on policies, programs, or products from the public, private, or non-for-profit sectors. Advocacy approaches were included. We excluded more clinically-oriented references to standards of practice.

Each researcher also listed the types of articles each journal publishes (e.g., research articles, policy briefs, or commentaries). Journals indicating a potential policy-related option in their types of articles were coded (yes or no) and the majority of information provided on this type of article was copied verbatim. Explicit instructions had to be provided relating to policy in defining the article option or the article explanation had to include policy-related language. We did not assume without explicit instruction that a type of article could serve as a policy-related option, including non-traditional research articles such as editorials or commentaries. Likewise, outlets for research, such as original research, research briefs, pooled or meta-analyses, or reviews were not coded unless explicit information describing these outlets were provided relating to policy. If listed within a journal's instructions to authors, position papers were coded as a potential option since the likelihood is high they will discuss policy; nonetheless, these statements could potentially only focus on setting forth professional standards and practices.

To further examine potential policy-related guidance, we focused on the article format for potential policy-related headings and guidance per article section. Specifically, two researchers determined the article format or section headers (e.g., introduction) used for each type of article option and documented information on each type of article's use of an abstract and, if structured, the headings used. Efforts were made in the few instances where this information was not provided in the guidelines to authors to review the formats used by articles published in the most recent journal issue. Journals indicating a potential policy-related outlet within an article structure were coded (yes or no) and any information provided to justify this code was copied verbatim. Unless explicit policy-related language was used in describing the section or heading, the following sections were not coded: Introduction, Methods, Results, Conclusion, or Comment. Although the structure of all article types were reviewed, we ultimately compared and contrasted each of the 50 journals' research structure, since this type of article was consistently found in all 50 journals. The scant guidance provided on structure on other types of articles was noted, if policy relevant, when coding the types of articles available by journal.

We tried to capture other forms of explicit guidance to authors about inclusion of policy-related research or implications in a particular article type or throughout the journal. Simple or vague references to "implications", "importance", "relevance", or even "recommendations" were not coded, as without further explanation these terms could be referring to implications or recommendations for future research. We also did not code "unqualified statements", "objective comments", or guidance emphasizing authors should avoid "making claims not supported by data." We did not code the term "practice" unless further policy-related language was used in defining this term.

Information provided on conflicts of interest, disclosing funding sources, transparency policies, or research ethics were not coded as policy for our purposes. We also excluded disclaimers, such as statements or opinions expressed in the journal do not reflect the journal, nor did we code a journal's stance on the National Institutes of Health's PubMed Repository or open access. We excluded stances on intellectual property and group authorship. Information on supplementary journals was only included if explicit policy-related information was mentioned.

Our coding protocol was pilot tested on the following five journals, not included in our list of 50: Canadian Journal of Public Health, New Zealand Medical Journal, Journal of Health Economics, Social Science & Medicine, and Hawaii Medical Journal. Inter-rater reliability for each step of determining the list of 50 journals and for all processes used in examining policy aims and instructions to authors was 100%. Minor reconciliations were made when in a few instances exclusions were miscoded.

Results

Table 1 lists in alphabetical order our final list of 50 journals, with information from the 5 approaches used to make this determination and each journal's category determination. Fifteen of the 50 journals (30%) included policy in their aims and scope (Table 2). Eighteen journals (36%) explicitly instructed authors on a type of article where policy-related information could be published (Table 3). Only three journals (6%) explicitly indicated a potential policy outlet within their research article format and one journal provided a potential policy outlet within their abstract structure (Table 4).

None of the examined journals covered policy-related information in all 5 areas (i.e., vision, guidance, outlet, structure, and abstract). The *British Medical Journal* and the *Journal of Nutrition Education and Behavior* covered the most areas (4 of the 5). Moreover, the *British Medical Journal* was the only journal coded for providing guidance beyond information provided in the vision, outlet, structure, or abstract areas: "We're simply aiming to make research articles so clear that peer reviewers, editors, clinicians, educators, ethicists, policy makers, systematic reviewers, guideline writers, journalists, patients, and the general public can tell what really happened during a study." The *American Journal of Health Promotion* had policy-related information in 3 of the 5 areas. Seven journals had policy-related information in 2 of the 5 areas: *Health and Place, Journal of the American Medical Association, Medical Journal of Australia, Ethnicity and Diseases, Journal of Adolescent Health, American Journal of Prevention Medicine*, and *American Journal of Public Health*. Thirteen journals had policy-related information in at least 1 area. In comparison to other leading childhood obesity journals, four of the journals covering at least one policy area published over 100 articles, according to our PubMed search: *American Journal of Clinical Nutrition, Journal of Clinical Endocrinology and Metabolism, Journal of Pediatrics*, and *Pediatrics*. All but one (*Circulation*, 14.595) of the five journals with an impact factor over 10 covered policy-related information (*British Medical Journal*, 12.827, *Journal of the American Medical Association*, 31.718, *Lancet*, 28.409, and *New England Journal of Medicine*, 50.017). Three of the top five pediatric journals to find the best evidence for clinical practice covered policy-related information (*Journal of Pediatrics*, 1st, *Pediatrics*, 3rd, and *Archives of Disease in Childhood*, 5th). Four of the top seven general internal medicine journals to find the best evidence for clinical practice covered policy-related information (*New England Journal of Medicine*, 1st, *Lancet*, 2nd, *Journal of the American Medical Association*, 4th, and *British Medical Journal*, 7th).

Table 1. Fifty Leading Childhood Obesity Journals Examined, Based on 5 Criteria

Listed in Alphabetical Order	Journal Name (Journal Category, Secondary Category)	Pub Med[1]	Web of Science[2]	ISI Web of Knowledge Impact Factor[3]	Garfield, et al.'s[4] Rankings for Journals with the: "Best Evidence for Clinical Practice"	NCI Food Environment Database[5]
1	Acta Paediatrica (Pediatrics)	88	6	1.517	NL	0
2	American Journal of Clinical Nutrition (Nutrition)	227	12	6.740	NL	1
3	American Journal of Epidemiology (Epidemiology)	35	4	5.454	NL	0
4	American Journal of Health Promotion (Public Health)	9	0	1.703*	NL	8
5	American Journal of Preventive Medicine (Public Health)	41	1	3.766	NL	14
6	American Journal of Public Health (Public Health)	38	0	4.241	NL	15
7	Appetite (Nutrition, Eating Behavior & Weight Regulation)	39	3	2.341	NL	2
Listed in	Journal Name	Pub	Web of	ISI Web of	Garfield, et al.'s[9] Rankings	NCI Food

---

[1] Number of articles published between 1998/01/01 through 2009/11/20.
[2] Number of articles published between 1998/01/01 through 2009/11/20.
[3] Impact factors for journals marked with an * were found on the journal's website.
[4] Garfield E, Parkin PC, & Birken C. In which journals will pediatricians find the best evidence for clinical practice? Pediatrics. 2000;106:377. Journals not listed in this article are marked as NL.
[5] Number of articles in database as of November 6, 2009.

Alphabetical Order	(Journal Category, Secondary Category)	Med[6]	Science[7]	Knowledge Impact Factor[8]	for Journals with the: "Best Evidence for Clinical Practice"	Environment Database[10]
8	Archives of Disease in Childhood (Pediatrics)	73	6	3.011	5th in Pediatrics	0
9	Archives of Pediatrics and Adolescent Medicine (Pediatrics)	93	4	4.320	NL	1
10	Asia Pacific Journal of Clinical Nutrition (Nutrition)	58	4	0.817	NL	0
11	BMC Public Health (Public Health)	30	9	2.029	NL	1
12	British Journal of Nutrition (Nutrition)	35	2	2.764	NL	0
13	British Medical Journal (Medicine)	98	2	12.827	7th in General and Internal Medicine	0
14	Circulation (Cardiovascular Diseases)	40	0	14.595	NL	0
15	Diabetes (Diabetes)	43	1	8.398	NL	0
16	Diabetes Care (Diabetes)	86	3	7.349	NL	0
17	Ethnicity and Diseases (Medicine, Health Disparities)	50	0	0.815	NL	1
18	European Journal of Clinical Nutrition (Nutrition)	108	3	2.686	NL	1
19	European Journal of Endocrinology (Endocrinology)	38	0	3.791	NL	0
20	European Journal of Pediatrics (Pediatrics)	46	4	1.416	NL	0
21	Health and Place (Geography)	5	3	2.818*	NL	4
22	Hormone Research (Endocrinology)	31	1	2.28	NL	0
23	International Journal of Behavioral Nutrition and Physical Activity (Nutrition)	0	8	2.32*	NL	6
24	International Journal of Eating Disorders (Psychology)	37	0	2.392	NL	0
25	International Journal of Epidemiology (Epidemiology)	22	0	5.838	NL	1

9 Garfield E, Parkin PC, & Birken C. In which journals will pediatricians find the best evidence for clinical practice? Pediatrics. 2000;106:377. Journals not listed in this article are marked as NL.

6 Number of articles published between 1998/01/01 through 2009/11/20.

7 Number of articles published between 1998/01/01 through 2009/11/20.

8 Impact factors for journals marked with an * were found on the journal's website.

10 Number of articles in database as of November 6, 2009.

Table 1. (continued)

Listed in Alphabetical Order	Journal Name (Journal Category, Secondary Category)	Pub Med[11]	Web of Science[12]	ISI Web of Knowledge Impact Factor[13]	Garfield, et al.'s[14] Rankings for Journals with the: "Best Evidence for Clinical Practice"	NCI Food Environment Database[15]
26	International Journal of Obesity[16] (Obesity, Nutrition)	219	17	3.640	NL	0
27	International Journal of Pediatric Obesity (Pediatrics)	85	0	3.984	NL	0
28	Journal of Adolescent Health (Pediatrics)	45	2	2.910	NL	1
29	Journal of the American Dietetic Association (Nutrition)	177	7	2.868	NL	12
30	Journal of the American Medical Association (Medicine)	41	0	31.718	4th in General and Internal Medicine	0
31	Journal of Clinical Endocrinology Metabolism (Endocrinology)	170	5	6.325	NL	0
32	Journal of Nutrition (Nutrition)	37	0	3.647	NL	0
33	Journal of Nutrition Education and Behavior (Nutrition)	21	4	1.743	NL	2
34	Journal of Pediatrics (Pediatrics)	143	3	4.122	1st in Pediatrics	0
35	Journal of Pediatric Endocrinology Metabolism (Pediatrics, Endocrinology)	107	8	0.938	NL	0
36	Journal of Pediatric Gastro-enterology and Nutrition (Pediatrics)	36	2	2.132	10th in Pediatrics	0
37	Journal of School Health (School Health)	17	4	1.273	NL	7
38	Lancet (Medicine)	58	1	28.409	2nd in General and Internal Medicine	0
39	Medical Journal of Australia (Medicine)	36	0	3.320	NL	0

11 Number of articles published between 1998/01/01 through 2009/11/20.
12 Number of articles published between 1998/01/01 through 2009/11/20.
13 Impact factors for journals marked with an * were found on the journal's website.
14 Garfield E, Parkin PC, & Birken C. In which journals will pediatricians find the best evidence for clinical practice? Pediatrics. 2000;106:377. Journals not listed in this article are marked as NL.
15 Number of articles in database as of November 6, 2009.
16 Journal Name as of 2005, previously International Journal of Obesity Related Metabolic Disorders, which published 375 articles in our PubMed search.

Listed in Alphabetical Order	Journal Name (Journal Category, Secondary Category)	Pub Med[17]	Web of Science[18]	ISI Web of Knowledge Impact Factor[19]	Garfield, et al.'s[20] Rankings for Journals with the: "Best Evidence for Clinical Practice"	NCI Food Environment Database[21]
40	Medicine Science in Sports and Exercise (Sports Medicine & Exercise Science)	28	1	3.399	NL	0
41	Metabolism (Diabetes)	57	0	2.920*	NL	0
42	New England Journal of Medicine (Medicine)	40	1	50.017	1st in General and Internal Medicine	0
43	Nutrition, Metabolism, & Cardiovascular Diseases (Cardiovascular Diseases, Nutrition)	32	0	3.565	NL	0
44	Obesity (Silver Spring)[22] (Obesity)	224	15	2.762	NL	0
45	Obesity Reviews (Obesity)	102	2	5.569	NL	0
46	Pediatric Diabetes (Pediatrics)	18	0	2.424	NL	0
47	Pediatric Research (Pediatrics)	28	0	2.604	2nd in Pediatrics	0
48	Pediatrics (Pediatrics)	178	19	4.789	3rd in Pediatrics	1
49	Preventive Medicine (Public Health)	74	5	2.757	NL	13
50	Public Health Nutrition (Nutrition)	108	15	2.123	NL	10

A number of trends emerged by journal category. Journals being categorized into the following categories all covered policy-related information: medicine (n=6), public health (n=5), geography (n=1), and school health (n=1). None of the journals categorized into the following categories covered policy-related information: cardiovascular diseases (n=2), diabetes (n=3), epidemiology (n=2), obesity (n=3), psychology (n=1), and sports medicine & exercise science (n=1). Our list of leading childhood obesity journals did not include any policy-oriented journals. The following list captures the journal categories with the highest percentage of policy areas covered (i.e., the number of policy-related areas covered per journal category/five studied areas times the number of journals characterized within each journal category): geography (40%), medicine (40%), public health (36%), school health (20%), pediatrics (12%), nutrition (12%), and endocrinology (7%).

17 Number of articles published between 1998/01/01 through 2009/11/20.
18 Number of articles published between 1998/01/01 through 2009/11/20.
19 Impact factors for journals marked with an * were found on the journal's website.
20 Garfield E, Parkin PC, & Birken C. In which journals will pediatricians find the best evidence for clinical practice? Pediatrics. 2000;106:377. Journals not listed in this article are marked as NL.
21 Number of articles in database as of November 6, 2009.
22 Formerly known as Obesity Research, which published 173 articles in our PubMed search.

Table 2. Policy Type Language in Journal's Scope, Mission, or Aim Statements (n=15 out of 50)[23]

Journal Name	(Section Policy Type Language was Found) Verbatim Policy Type Language Coded (Underlining added for emphasis)
American Journal of Health Promotion	(Merging Science and Practice): Providing a forum for discussion among diverse disciplines. Advancing our field is further hampered by fragmentation caused by the many disciplines involved in health promotion. For example, in designing a weight control program a nutritionist might focus on the best food to eat, a physiologist on the amount and type of exercise a psychologist on the connections between eating and emotions, a physician on metabolic imbalances, a social worker on work and family factors, an anthropologist on the influence of societal norms, a teacher on curriculum exposure, a communication specialist on advertising and media influences, a political scientist on agricultural subsidy programs, a city planner on access to sidewalks and parks, and a transportation engineer on access to public transit an economist on the relative cost of healthy versus unhealthy foods. The best solution would be to include input from all of these and other disciplines. Ultimately, to help the full population achieve optimal health, we need to create communities that provide all residents with the knowledge, encouragement, opportunity and resources necessary for ongoing healthy lifestyle practices. To achieve these goals, we need to draw on the skills of traditional public health, medical and clinical fields as well as those who plan and design our communities and the laws and policies that govern them. To meet this goal of integrating multiple disciplines The *American Journal of Health Promotion* strives to engage diverse disciplines through our broad and diverse editorial content and editorial board. Our 18 editorial sections include four broad categories: Interventions, Strategies, Applications and Research Methods. Original research, literature reviews and editorials Our reviewers include over 250 of the top scientists and practitioners drawn from diverse fields. (Editorial Scope): Strategies, including awareness programs, behavior change programs, and supportive environment programs such as culture change, health policy, and population health
American Journal of Preventive Medicine	(Aims and Scope): The *American Journal of Preventive Medicine* is the official journal of the American College of Preventive Medicine and the Association for Prevention Teaching and Research. It publishes articles in the areas of prevention research, teaching, practice and policy. Original research is published on interventions aimed at the prevention of chronic and acute disease and the promotion of individual and community health. Of particular emphasis are papers that address the primary and secondary prevention of important clinical, behavioral and public health issues such as injury and violence, infectious disease, women's health, smoking, sedentary behaviors and physical activity, nutrition, diabetes, obesity, and alcohol and drug abuse. Papers also address educational initiatives aimed at improving the ability of health professionals to provide effective clinical prevention and public health services. Papers on health services research pertinent to prevention and public health are also published. The journal also publishes official policy statements from the two co-sponsoring organizations, review articles, media reviews, and editorials. Finally, the journal periodically publishes supplements and special theme issues devoted to areas of current interest to the prevention community.
American Journal of Public Health	(Mission): Promoting public health, policy, practice, and education is the foremost mission of the American Journal of Public Health. We aim to embrace all of public health, from global policies to the local needs of public health practitioners. Contributions of original unpublished research and practice articles and briefs, social science and policy analyses (analytic essays), constructively critical commentaries, department papers, and letters to the editor and responses are welcome.
Archives of Disease in Childhood	(Aims and Scope): Archives of Disease in Childhood is an international peer review journal that aims to keep pediatricians and others up to date with advances in the diagnosis and treatment of childhood diseases as well as advocacy issues such as child protection. It focuses on all aspects of child health and disease from the perinatal period (in the Fetal and Neonatal edition) through to adolescence. ADC includes original research reports, commentaries, reviews of clinical and policy issues, and evidence reports. Areas covered include: community child health, public health, epidemiology, acute pediatrics, advocacy, and ethics. New sections include: guidelines update; international health; a column written by patients about their experience with the health care system; and abstracts from Journal Watch Pediatrics and Adolescent Medicine.

---

23 All references were extracted from the respective journal's online aims and scope and/or instructions to authors.

Journal Name	(Section Policy Type Language was Found) Verbatim Policy Type Language Coded (Underlining added for emphasis)
Archives of Pediatrics and Adolescent Medicine	**(Mission Statement):** The *Archives of Pediatrics & Adolescent Medicine* is a monthly, peer-reviewed journal for physicians and other health care professionals who contribute to the health of children and adolescents. The *Archives of Pediatrics & Adolescent Medicine* provides an open forum for dialogue on a full range of clinical, scientific, advocacy, and humanistic issues relevant to the care of pediatric patients from infancy through young adulthood. The *Archives of Pediatrics & Adolescent Medicine* is a vehicle for increased attention to adolescent health, the education of pediatric health care professionals, and disease prevention and health promotion. The *Archives of Pediatrics & Adolescent Medicine* publishes original studies, editorials, reviews by experts, practice commentaries, case quizzes, and updates on clinical science and practice management. The Pediatric Forum provides our readers with opportunities to express their views. The *Archives of Pediatrics & Adolescent Medicine* incorporates *AJDC*, the oldest journal in US pediatric literature, which originated in 1911.
BMC Public Health	(Scope): *BMC Public Health* is an Open Access, peer-reviewed journal that considers articles on the epidemiology of disease and the understanding of all aspects of public health. The journal has a special focus on the social determinants of health, the environmental, behavioral, and occupational correlates of health and disease, and the impact of health policies, practices and interventions on the community.
British Medical Journal	(About BMJ): All the *BMJ*'s original research is published in full on bmj.com, with open access and no limits on word counts. We do not charge authors or readers for research articles, nor for other articles arising from work funded by open access grants. The *BMJ*'s vision is to be the world's most influential and widely read medical journal. Our mission is to lead the debate on health and to engage, inform, and stimulate doctors, researchers, and other health professionals in ways that will improve outcomes for patients. We aim to help doctors to make better decisions. The BMJ team is based mainly in London, although we also have editors elsewhere in Europe and in the US.
Ethnicity and Disease	(Introduction): The journal also provides information in special sections dedicated to legislative and regulatory issues...
Health and Place	The journal is an interdisciplinary journal dedicated to the study of all aspects of health and health care in which place or location matters. Recent years have seen closer links evolving between medical geography, medical sociology, health policy, public health and epidemiology. The journal reflects these convergences, which emphasize differences in health and health care between places, the experience of health and care in specific places, the development of health care for places, and the methodologies and theories underpinning the study of these issues. The journal brings together international contributors from geography, sociology, social policy and public health. It offers readers comparative perspectives on the difference that place makes to the incidence of ill-health, the structuring of health-related behavior, the provision and use of health services, and the development of health policy. At a time when health matters are the subject of ever-increasing attention, *Health & Place* provides accessible and readable papers summarizing developments and reporting the latest research findings.
International Journal of Behavioral Nutrition and Physical Activity	*IJBNPA* considers manuscripts in the following areas: Behavioral interventions (including community-based and clinical) Population behaviors (including measurement, prevalence and etiology) Predictors of behavior Innovative behavioral theories Measurement issues Policy and public health issues
International Journal of Pediatric Obesity	(Aims and Scope): The *International Journal of Pediatric Obesity* is a new, peer-reviewed, quarterly journal devoted to research into obesity during childhood and adolescence. The topic is currently at the centre of intense interest in the scientific community, and is of increasing concern to health policy-makers and the public at large. Areas of research, include: Community and public health measures to prevent overweight and obesity in children and adolescents
Journal of Adolescent Health	(About the Journal of Adolescent Health): *The Journal of Adolescent Health* is a multidisciplinary scientific Journal, which seeks to publish new research findings in the field of Adolescent Medicine and Health ranging from the basic biological and behavioral sciences to public health and policy. We seek original manuscripts, review articles, letters to the editor, commentaries, and case reports from our colleagues in Anthropology, Dentistry and Oral Health, Education, Health Services Research, International Health, Law, Medicine, Mental Health, Nursing, Nutrition, Psychology, Public Health and Policy, Social Work, Sociology, Youth Development, and other disciplines that work with or are committed to improving the lives of adolescents and young adults.

Table 2. (continued)

Journal Name	(Section Policy Type Language was Found) Verbatim Policy Type Language Coded (Underlining added for emphasis)
Journal of American Medical Association	(Key Objectives, # 8) To inform readers about nonclinical aspects of medicine and public health, including the political, philosophic, ethical, legal, environmental, economic, historical, and cultural
Journal of Nutrition Education and Behavior	(Author Guidelines): The *Journal of Nutrition Education and Behavior* welcomes evidence-based manuscripts that provide new insights and useful findings related to nutrition education research, practice, and policy.
Medical Journal of Australia	(Advice to Authors): The *Medical Journal of Australia* aims to be the premier forum for information and commentary on clinical medicine and healthcare in Australia. To achieve this, the *MJA* publishes original clinical research, reviews and educational articles, together with commentary and informed debate on standards of clinical practice, and on social, ethical, legal and other issues related to healthcare. The Journal welcomes articles in all these categories.

Table 3. Journals Indicating Potential Policy-Related Published Options in their Types of Articles (n=18 out of 50)[1]

Journal Name	Verbatim Policy Type Language Coded (Underlining added for emphasis)
American Journal of Clinical Nutrition	Invited controversy position papers or special invited articles including controversies and perspectives
American Journal of Health Promotion	Editorial Content Areas: Strategies: Health Policy
American Journal of Preventive Medicine	Mission Statement, not list of types of articles: Publishes official policy statements from the two co-sponsoring organizations
American Journal of Public Health	Commentaries (included policy manuscripts) Analytic Essays (designed specifically to meet the needs of Health Policy and Ethics Forum, Government, Politics, and Law, and Framing Health Matters. ...Created to provide a forum for critical analyses of public health issues from disciplines other than biomedical sciences, including but not limited to social sciences, human rights, political science, and ethics. Included examples. Government, politics, and law: Department Editor believes a useful approach in public health for translating research findings into everyday practice relevant for our readers "might involve soliciting the work of leading public health system researchers as well as other scientists, and encouraging translation of their findings to the public health practitioner at all levels. There are also extremely important transforming events taking place in the larger public health field. Stated more succinctly, the remodeled Government, Politics, and Law department will emphasize the metamorphosis that public health practice is undergoing—focusing on what it can become." Utilizes the analytic essay.
British Medical Journal	We publish original research articles, review and educational articles, letters, and articles commenting on the clinical, scientific, social, political, and economic factors affecting health. Quality improvement reports included strategies for change, effects of change, and lessons learnt.
Ethnicity and Disease	(Introduction, not Types of Articles): The journal also provides information in special sections dedicated to legislative and regulatory issues...
Health and Place	Review papers may provide scholarly assessments of new policies or practices, or academic overviews of new areas of study (5000-6000 words).
Journal of Adolescent Health	Review Articles: Authors need to identify the content area for the review article that they are submitting: Behavioral/Psychosocial Sciences, Clinical/Health Sciences, Public Policy/Public Health, International Health, or Youth Development.

---

[1] All references were extracted from the respective journal's online instructions to authors.

Journal Name	Verbatim Policy Type Language Coded (Underlining added for emphasis)
Journal of American Medical Association	Commentary. These papers may address virtually any important topic in medicine, public health, research, ethics, health policy, or health law and generally are not linked to a specific article. Commentaries should be well focused, scholarly, and clearly presented and must have no more than 2 authors. Maximum length: up to 1200 words of text—or 1000 words of text with 1 small table or figure—and no more than 10 references. Commentaries not meeting these guidelines will not be considered. Controversies. The Controversies section provides a forum for debate, discussion, and examination of challenging issues and difficult questions in clinical medicine, research, public health, or health policy. The section uses a "view-counterview" or "point-counterpoint" format, in which 2 articles address the topic or issue and present opposing viewpoints or positions. Manuscripts should be well argued, cogent, and scholarly, with appropriate documentation for scientific content and factual assertions and should have no more than 2 authors. Recommended length: up to 1200 words of text, with no more than 10 references and 1 table or figure. Authors interested in submitting a paper for the Controversies section should contact the editorial office prior to manuscript submission by sending an e-mail to phil.fontanarosa@jama-archives.org. Health Law and Ethics. These manuscripts discuss important issues in law and/or ethics with implications for patient care or physician practice and are of general medical interest. The manuscript may report original data from a research investigation or it may be a scholarly, well-referenced, systematic review of a topic. Systematic reviews should include a Methods section with descriptions of the data search and selection process, critical evaluation of the cited articles or data sources, and acknowledgment of any controversial aspects of the topic. The manuscript content must be original, and manuscripts derived from previously published material (for example, a law review article) are not acceptable. For reports of original data and systematic reviews, a structured abstract is required; see instructions for preparing structured abstracts. Maximum length: up to 3000 words of text (not including references or abstract) with no more than a total of 4 tables and/or figures and no more than 50 references. Manuscripts submitted for Health Law and Ethics may be recommended for consideration as Commentaries after favorable peer review. Authors interested in submitting a Health Law and Ethics manuscript should contact the editorial office prior to manuscript preparation and submission by sending an e-mail to helene.cole@jama-archives.org. Special Communication. These manuscripts describe an important issue in clinical medicine, public health, health policy, or medical research in a scholarly, thorough, well-referenced, systematic or evidence-based manner. A narrative (unstructured) abstract of 200 words or fewer is required. Maximum length: 3000 words of text (not including tables, figures, or references) with no more than a total of 4 tables and/or figures and no more than 50 references. Authors interested in submitting a Special Communication manuscript must contact the editorial office prior to manuscript preparation and submission by sending an e-mail to cathy.deangelis@jama-archives.org.
Journal of Clinical Endocrinology and Metabolism	Position and Consensus Statements related to the endocrine and metabolic health standards and healthcare practices may be submitted by professional societies, task forces, and other consortia.
Journal of Nutrition Education and Behavior	*Reports* are articles that (1) discuss policy issues relevant to nutrition education and behavior [outlet] or (2) review emerging topics as they relate to nutrition education and behavior.
Journal of Pediatrics	Invited Commentaries: Commentaries are generally invited only. Authors who wish to propose a Commentary should e-mail a proposal letter and outline to the Editors for approval *before* submitting the full manuscript. Commentaries should serve as a forum for governmental health policies, economic issues, medical/scientific ethics, psychosocial issues, and international health, particularly in the developed world.

Table 3. (contined)

Journal Name	Verbatim Policy Type Language Coded (Underlining added for emphasis)
Journal of School Health	Commentaries include position papers, analyses of current or controversial issues, and creative, insightful, reflective treatments of topics related to healthy youth pre-K to 12th grade or healthy school environments. School Health Policy: Health Policy articles present an interdisciplinary analysis of school health policies affecting healthy youth pre-K to 12th grade or healthy school environments. Manuscripts should focus on new scientific results and policy reviews that shed light on important debates and controversies. They should provide insightful, thought-provoking examinations of school policies and analyses of controversial policy issues that have the potential to affect the health or safety of students or school staff. The focus on policy may be from a policy making, policy-implementation, or policy-impact perspective. School Health Policy articles require a structured abstract (see more details below under Abstract). For further clarification/information regarding the Health Policy column, contact the Associate Editor, Sherry Everett Jones. Legal Issues in School Health: Legal Issues articles critically analyze legislation, regulations, or judicial rulings that have the potential to affect the health, mental health, or safety of pre-K to 12th grade students or school staff; provide creative discussion of current and emerging legal issues on school health-related topics, including legal issues related to the health of children and adolescents in schools; describe results of empirical research on legal issues; and /or offer critical dialogue on legal topics which impact coordinated school health programs. Legal Issues articles require a structured abstract (see more details below under Abstract). For further clarification/information regarding the Legal Issues column, contact the Associate Editor, Sherry Everett Jones.
Lancet	Articles: Global public-health and health-policy research are other areas of interest to *The Lancet*
Medical Journal of Australia	Original clinical research, reviews and educational articles, together with commentary and informed debate on standards of clinical practice, and on social, ethical, legal and other issues related to healthcare. Medicine and the Law: The interface between medicine and law Position Statement: Recommendations from learned bodies or interest groups [not explicitly policy] Viewpoint: Any issues related to healthcare/ethics/law as applicable to medicine
New England Journal of Medicine	Special articles are scientific reports of original research in such areas as economic policy, ethics, law, and health care delivery. Health Law, Ethics, and Human Rights are nearly always solicited, but we are willing to consider unsolicited manuscripts or proposals for manuscripts. Please contact the Editorial Office before submitting a manuscript. Health Policy Reports are nearly always solicited, but we are willing to consider unsolicited manuscripts or proposals for manuscripts. Please contact the Editorial Office before submitting a manuscript.
Pediatrics	Note http://aappolicy.aappublications.org/ is where to find policy statements
Preventive Medicine	Special Issues or Sections Devoted to Specific Topics of Significance Have Included: Prevention of cervical cancer: HPV vaccines, screening, and other policy options

Table 4. Potential Policy Outlets within the Research Article Structure (n=3 out of 50)[1]

Journal Name	Section Header Coded: Verbatim Policy Type Language Coded (Underlining added for emphasis)
American Journal of Health Promotion	SO WHAT: Consider…What are the implications for health promotion practice or research? [Not explicit on policy, but outlet different from traditional Introduction, Methods, Results, Discussion format and prior research indicates policy comments often occurred in this section]
British Medical Journal	Structured discussion: Structure includes five paragraphs, the fourth paragraph discusses the meaning of the study: possible explanations and implications for clinicians and policymakers. "What this paper adds" box: Please produce a box offering a thumbnail sketch of what your article adds to the literature, for readers who would like an overview without reading the whole article. …. You might use the last sentence to summarize any implications for practice, research, policy, or public health.
Journal of Nutrition Education and Behavior	Structured Abstract's Conclusions and Implications: Specifies study conclusions directly supported by results reported in the abstract and specifies implications for research, practice, and policy making. Implications for Research and Practice: Specifies how researchers, practitioners, and policy makers could apply results to future work.

Discussion

Our study purpose was to examine leading childhood obesity journals' instructions to authors regarding policy-related research and implications in order to inform actionable guidance on how journals may advance the formation and evaluation of evidence-based obesity policy. Despite increasing calls by researchers for policy to be further developed as a critical component of a multilevel childhood obesity prevention strategy(20, 24), few of the journals examined for this study had policy-related aims or policy-relevant guidance for authors. This finding confirmed our hypotheses. Above all, the significance of these findings is the systematic illustration of the current situation from vision to instruction, abstract to discussion.

More than a quarter of the studied journals (30%) included policy in their aims and scope, documenting interest in shaping the debate on relevant policies. The *British Medical Journal* stated: "Our mission is to lead the debate on health and to engage, inform, and stimulate doctors, researchers, and other health professionals in ways that will improve outcomes for patients." The *Journal of American Medical Association* set forth a more modest objective:

"To inform readers about nonclinical aspects of medicine and public health, including the political, philosophic, ethical, legal, environmental, economic, historical, and cultural." According to our analyses, eight journals (*American Journal of Clinical Nutrition, Journal of Clinical Endocrinology Metabolism, Journal of Pediatrics, Journal of School Health, Lancet, New England Journal of Medicine, Pediatrics*, and *Preventive Medicine*) did not express any policy-related language in their aims and scope while still offering potential policy-related instruction elsewhere. On the other hand, five journals (*Archives of Disease in Childhood, Archives of Pediatrics and Adolescent Medicine, BMC Public Health, International Journal of Behavioral Nutrition and Physical Activity*, and *International Journal of Pediatric Obesity*) expressed a policy vision but provided no further detail or instruction within their author

---

[1] All references were extracted from the respective journal's online instructions to authors.

guidelines demonstrating where and how this vision may be carried out throughout the journal. Whether or not a journal explicitly aims to inform or advance evidence-based policy or includes in their scope publishing articles in the areas of public health policy is a journal's decision. No data exists to contend focusing a journal's aim to include policy has more effective impact on translating science into policy. The point of this analysis was to illustrate if and, to what extent, leading childhood obesity journals were specifically envisioning their role in the evidence-based policy process, as well as aiming to be a forum for policy discourse. Our findings may stimulate not only these evaluated journals but any type of journal to take pause to evaluate how they address policy in their aims and scope. Does the absence have any implications? Is a possible disconnect between a journal's vision and the other guidance to authors provided important to find?

Over a third (36%) of the evaluated journals specified types of articles where policy-related information could be published. Even more clinically-oriented journals (e.g., *American Journal of Clinical Nutrition*) and medical journals (e.g., *British Medical Journal*) offered a policy-related article option. In fact, all journals categorized into the medicine category discussed at least one policy component. Only three journals (6%) explicitly indicated a potential policy outlet within their research article format and one journal provided a potential policy outlet within their abstract structure. An analysis published in 1999 found 24% of a sample of epidemiology articles contained policy statements(25). These findings, while dated, illustrate how journals and authors might not be aware of the extent policy statements are discussed in their publications. Instructions to authors may be one possible venue for journals to communicate their stance on whether or not policy should be included in submissions and, if so, what types of publications work best for this discussion, where within the article could these discussions take place, and how best to integrate policy into a study design and articulate objective and informative policy implications.

Epidemiology has a stance against authors including policy implications in their research article(5). In our study, only two epidemiology journals were examined and neither addressed policy—whether it should or should not be included. We recognize arguments and rationale for not including policy implications in research articles, particularly when a word count limits authors to delineate likely complex implications into one or two sentences followed by "this study informs policy." Future efforts at the level of a journal's board of editors or collective endeavors amongst multiple editors-in-chief might consider devising more explicit editorial policies and best practices on how researchers could incorporate policy components in their study design and articulate policy implications in their research article or other communication venues. Given calls for more methodological rigor in obesity policy research(20), journals might expand their guidance beyond policy implications and consider providing guidance on how to increase sophistication in study design, measurements, and statistical approaches for studies focusing on policy.

Both researchers and policymakers often express sentiments like "policymakers don't read research articles"(26). This study only documents if a journal listed a potential type of article an author could consider for submitting an article relating to public health policy. Future research may retrospectively examine the quantity of policy-related articles being published, noting the number of articles published in the various article type categories. This could be narrowed to only focus on obesity or to examine if and how obesity policy discourse compares to other policy-related discussions, such as tobacco control or access to health care. Additional research could investigate the extent existing discourse on policy in journal articles influences the translation of evidence into policy, taking into consideration the affects may be different for different stakeholders. Since no policy-oriented journals were found to be within the leading childhood obesity journals, policy discourse on childhood obesity might not be published in the most read, highest impact journals. Future work might explore if and how the lack of policy discourse within leading childhood obesity journals affects translation.

Our study was limited in scope to leading childhood obesity journals. Only online instructions were reviewed, possibly missing instructions provided in editorials or other formats. A journal's vision and guidelines to authors may not reflect what the journal actually publishes. In spite of these limitations, this systematic approach to examining leading childhood obesity journals' instructions to authors regarding policy-related research and implications offers intriguing data on the opportunities and gaps within author guidelines. These insights are particularly useful to consider in a time where increasing attention is given by both researchers and policymakers to the role of policy strategies in obesity prevention and treatment, along with a simultaneous demand for ensuring the translation of objective evidence-based research into policy.

Acknowledgments

Robin McKinnon, PhD, MPA, Risk Factor Monitoring and Methods Branch of the Applied Research Program at the National Cancer Institute of the U.S. National Institutes of Health provided us with invaluable feedback throughout the study design and manuscript preparation.

The University of North Carolina-Chapel Hill Health Sciences Librarians were of great assistance to us. This project is supported by the National Institute of Health (NIH) University of North Carolina Interdisciplinary Obesity Training Grant (T 32 MH75854-03). The content is solely the responsibility of the authors and does not necessarily represent the official views of the NIH.

References

[1]   Ogden, C., Carroll, M., Curtin, L., Lamb, M., and Flegal, K. "Prevalance of high body mass index in US children and adolescents, 2007-2008," *JAMA* 303 (2010).

[2]   Wojcicki, J., and Heyman, M. "Let's Move--Childhood obesity prevention from pregnancy and infancy onward," *N Engl J Med* 2010 (2010).

[3]   Teret, S. "So what?," *Epidemiology* 4 (1993): 93-94.

[4]   Rothman, K. "Policy recommendations in epidemiology research papers," *Epidemiology* 4 (1993): 94-95.

[5]   Editor-in-Chief. "Our policy on policy," *Epidemiology* 12 (2001): 371-372.

[6]   Coughlin, S. "A ban on policy recommendations in epidemiology research papers? Surely, you jest!," *Epidemiology* 5 (1993): 257-258.

[7]   Savitz, D., Poole, C., and Miller, C. "Reassessing the role of epidemiology in public health," *Am J Public Health* 89 (1999): 1158-1161.

[8]   Rothman, K., and Poole, C. "Science and policy making," *Am J Public Health* 75 (1985): 340-341.

[9]   Samet, J., and Lee, N. "Bridging the gap: Perspectives on translating epidemiologic evidence into policy " *Am J Epidemiology* 154 (2001): S1-S3.

[10]   Marks, J. "Epidemiology, public health, and public policy," *Prev Chronic Dis* 6 (2009): 1-4.

[11]   Stoto, M., Hermalin, A., Rose, L., Martin, L., Wallace, R., and Weed, D. "Advocacy in epidemiology and demography," *Ann N Y Academy Sci*: 76-87.

[12]   Glasgow, R., and Emmons, K. "How can we increase translation of research into practice? Types of evidence needed," *Annu Rev Public Health* 28 (2007): 413-433.

[13]   Samet, J. "Epidemiology and policy: The pump handle meets the new millennium," *Epidemiologic Reviews* 22 (2000): 145-154.

[14]   Oliver, T. "The politics of public health policy," *Annu Rev Public Health* 27 (2006): 195-233.

[15]   Green, L., Glasgow, R., Arkins, D., and Stange, K. "Making evidence from research more relevant, useful, and actionable in policy, program planning, and practice: Slips "Twixt Cup and Lip"," *Am J Prev Med* 37 (2009): S187-S191.

[16]   Briss PA, Brownson, R., Fielding, J., and Zaza, S. "Developing and using the Guide to Community Preventive Services: Lessons learned about evidence-based public health," *Annu Rev Public Health* 25 (2004): 281-302.

[17]   "Brass CT, Nunez-Neto B,& Williams ED. Congressional Research Service Report for Congress: Congress and Program Evaluation: An Overview of Randomized Controlled Trials (RCTs) and Related Issues. March 7, 2006.."

[18]   "Institute of Medicine, Committee on an Evidence Framework for Obesity Prevention Decision Making. Bridging the Evidence Gap in Obesity Prevention: A Framework to Inform Decision Making. Washington, DC: National Academics Press; 2010."

[19]   Huang, T., and Glass, T. "Transforming research strategies for understanding and preventing obesity," *JAMA* 300 (2008): 1811-1813.

[20]   Sallis, J., Story, M., and Lou, D. "Study designs and analytic strategies for environmental and policy research on obesity, physical activity, and diet: recommendations from a meeting of experts " *Am J Prev Med* 36 (2009): S72-S77.

[21]   Story, M., Sallies, J., and Orleans, T. "Editorial: Adolescent obesity: Towards evidence-based policy and environmental solutions," *J Adolescent Health* 45 (2009): S1-S5.

[22]   Story, M., Kaphingst, K., Robinson-O'Brien, R., and Glanz, K. "Creating healthy food and eating environments: Policy and environmental approaches.," *Annu Rev Public Health* 29 (2008): 253-272.

[23]   Garfield, E., Parkin, P., and Birken, C. "In which journals will pediatricians find the best evidence for clinical practice?," *Pediatrics* 106 (2000): 377.

[24]   "Institute of Medicine, Committee on Prevention of Obesity in Children and Youth, Koplan J, et al. *Preventing Childhood Obesity: Health in the Balance.* Washington, DC: National Academies Press; 2005."

[25]   Jackson, L., Lee, N., and Samet, J. "Frequency of policy recommendations in epidemiologic publications," *Am J Public Health* 89 (1999): 1206-1211.

[26]   Brownson, R., Chriqui, J., and Stamatakis, K. "Understanding evidence-based public health policy," *Am J Public Health* 99 (2009): 1-8.

In: Childhood Obesity: Risk Factors, Health Effects…   ISBN: 978-1-61761-982-3
Editor: Carol M. Segel   ©2011 Nova Science Publishers, Inc.

Chapter XI

Nutrition Education in School-Based Childhood and Adolescent Obesity Prevention Programs

Melinda J. Ickes [*] *and Manoj Sharma* [≠]
University of Cincinnati, Cincinnati, OH, USA

Abstract

The purpose of this chapter was to review nutrition education in school-based childhood and adolescent obesity prevention programs. Problems of overweight and obesity have reached epidemic proportions in children and adolescents. With the large number of children and adolescents affected by overweight and obesity, it has been recognized as a critical public health threat, and prevention should become a priority. Schools have been considered an important venue for the delivery of health education programs since a majority of the nation's youth can be reached. Implementing a program in the school does not necessarily correspond to success, and therefore common components among successful nutrition education programs were examined. Providing age-appropriate, culturally sensitive instruction and materials were connected to the success of the programs. Success was also found when the nutrition education program was relevant to the student, and information was applicable to their everyday life. It was also critical to go beyond dissemination of knowledge and develop attitudes, skills, and behaviors to adopt, maintain, and enjoy healthy nutrition habits. Finally, recommendations were made to enhance the effectiveness of future school-based nutrition education programs.

[*] Corresponding Author: (859) 572-5196 (Phone),(513) 556-3898 (Fax), ickesmj@uc.edu (e-mail)
[≠] Corresponding Author: (513) 556-3878 (Phone), (513) 556-3898 (Fax), manoj.sharma@uc.edu (e-mail)

Introduction

Prevalence

Problems of overweight and obesity have reached epidemic proportions (Ogden, Flegal, Carroll, & Johnson, 2002). Globally, approximately 1.6 billion adults (age 15+) were classified as overweight and of those at least 400 million were obese (World Health Organization, 2005). The World Health Organization projects these statistics to increase; by 2015, approximately 2.3 billion adults will be overweight and more than 700 million will be obese all over the world. In the United States, obesity prevalence doubled among adults between 1980 and 2004. More than one-third of adults, or over 72 million people, were obese in 2005-2006, including 33.3% among adult men and 33.2% among adult women (U.S. Department of Health and Human Services, 2007).

Unfortunately, the obesity epidemic has not been limited to adults but the percentage of children and adolescents who were overweight has also continued to increase. Obesity has been one of the most widespread health threats facing children and adolescents (Evans, Finkelstein, Kamerow & Renaud, 2005). Many health professionals have considered it to be a serious health issue which has placed a burden on our society. In the United States, childhood obesity was defined as being at or above the 95th percentile according to the Centers for Disease Control and Prevention (CDC) age-and gender-specific body mass index charts. Those considered at-risk for overweight were at or above the 85th percentile and below the 95th percentile (CDC, 2005).

According to the 2005-2006 National Health and Nutrition Examination Survey (NHANES) data, among 2-19 year old children, 17.1% were considered overweight: 13.9% of 2-5 year old, 18.8% of 6-11 year old, and 17.4% of 12-19 year old children (CDC, 2005). Variances exist among the states, with seven states incurring obesity rates of 15% - 19%. Among children and teenagers aged 6 - 19, 16% (almost 9 million) were overweight (CDC, 2005; Kaufman, 2003; USDHHS, 2004, 2007). This suggested that overall, 10% of the school-aged children in the world were overweight (Lobstein, Baur, Uauy, & IASO International Obesity Task Force, 2004). There has been a strong link between obesity during childhood and adolescence into adulthood, making the prevention of obesity in these populations a priority (Wilson, 2008).

Consequences of Child and Adolescent Obesity

Child and adolescent obesity has been of concern to many health professionals, parents, teachers, and school administrators because of the negative impact on the child's physical, psychological, behavioral, and social health. See Table 1 for a summary of these consequences. Children who were overweight had five times the risk of low health-related quality of life compared to children of normal weight (Schwimmer, Burwinkle, & Varni, 2003).

Physical Consequences. The dangers of being overweight or obese in childhood and adolescence have been well established. The Bogalusa Heart Study conducted in the United

States found that by 10 years of age, 60% of overweight children had at least one biochemical or clinical cardiovascular risk factor and 25% had more than two (Freedman, Dietz, Srinivasan, & Berenson, 1999). Additional long-term health consequences of childhood overweight included an increased risk of developing high blood pressure, high cholesterol, type 2 diabetes, stroke, certain types of cancer, gallbladder disease, arthritis, sleep disturbances, and asthma (USDHHS, 2007). Reportedly, physicians were increasingly treating pediatric patients who displayed symptoms of chronic diseases previously associated primarily with adulthood. Children were displaying conditions associated with serious health threats and a decreased life span (Shelton, 2008).

In addition, those who were overweight during childhood were more likely to maintain obesity into adulthood (Whitaker, Wright, Pepe, Seidel, & Dietz, 1997). In fact, a comprehensive review of the literature on childhood overweight persisting into adulthood indicated that 42% - 63% of overweight school-age children became obese adults, and the risk of obesity in adulthood was at least twice as high for overweight children as for non-overweight children (Smith, 2004). For the first time in history, the steady increase in life expectancy was threatened by a generation who would more than likely live unhealthier and shorter lives than their parents (Wilson, 2008).

Psychological Consequences. Children and adolescents who were overweight also experienced a variety of psychosocial issues: depression, peer rejection, low self-esteem, stigmatization, and discrimination (Erickson, Robinson, Haydel, & Killen, 2000; Schwartz & Puhl, 2003; Strauss, 2000). A widespread stereotype and negative attitude towards overweight and obese individuals has led to discrimination, as well as feelings of shame and guilt amongst these individuals (Cossrow, Jeffrey, & McGuire, 2001). Only a few studies have been reported on the topic of overweight stigmatization, but common threads of prejudicial treatment and discrimination were apparent. This stigmatization led to the internalization of negative self-evaluations which typically resulted in self-rejection. According to Charles Horton Cooley's (1964) looking-glass self theory, we see ourselves as others see us. Thus, children dealing with weight issues were exposed to a negative environment, which tended to be detrimental to their psychological and social functioning.

Societal Consequences. The effects of obesity on society were also indicated, primarily by imposing a significant financial burden on the health-care system. In the United States, the national health care expenditures related to overweight and obesity in adults were estimated between $98 billion and $129 billion (Institute of Medicine, 2004). Over the past two decades, costs associated with obesity-related illnesses specifically in children rose from $35 million to $127 million. States' annual medical spending for prevention, diagnosis, and treatment of individuals with obesity was estimated to range from $87 million to $7.7 billion across various states (Pyle, Sharkey, Yetter, Felix, & Furlong, 2006). Medical costs associated with overweight and obesity involved both direct and indirect costs. According to Finkelstein, Fiebelkorn and Wang (2003), direct medical costs included preventive, diagnostic, and treatment services. In contrast, indirect costs were related to morbidity (income lost from decreased productivity, restricted activity, and absenteeism) and mortality costs (the value of future income lost by premature death). There was a 197% increase in obesity-related hospitalizations among adolescents in the past 20 years, which has had a dire effect on the health care costs within this population (World Health Organization, 2003). The economic costs have also reached the school systems, where burdens related to students' absences cost districts between $9 and $20 per student (Action for Healthy Kids (AFHK),

2004). Since several states' funding was determined by attendance rates, the costs could be two-fold and should be considered a priority.

Table 1. Consequences of child and adolescent obesity

Types of consequences	
Physical	*Increased risk of developing:* Cardiovascular disease High blood pressure High cholesterol Type II diabetes Stroke Certain types of cancer Gallbladder disease Arthritis Sleep disturbances Asthma Maintain obesity into adulthood
Psychological	Depression Peer rejection Low self-esteem Feelings of shame and guilt Stigmatization Discrimination
Societal	Medical costs Morbidity costs Mortality costs School costs

Determinants of Overweight and Obesity

The factors which have contributed to overweight and obesity were considered numerous and complex. Body weight has been shaped by a combination of genetic, metabolic, behavioral, environmental, cultural, and socioeconomic influences. Some of the determinants are non-modifiable such as genetics, sex, age, and race. However, most of the population was more likely affected by the influences of surroundings, opportunities, or conditions predisposing individuals to obesity (Moreno et al., 2004).

The reason most often cited to cause obesity was related to an excess in energy consumption and/or inadequate physical activity (Office of the Surgeon General, 2001). Unhealthy eating and physical activity practices that contribute to childhood obesity tend to be established early in life; young persons having these unhealthy habits were likely to maintain them as they aged (CDC, 2005). Supported by the American Medical Association (Goutham, 2008), there were four well-recognized behaviors that have the capability of offsetting the development of childhood obesity (Table 2).

Therefore, when considering what will evoke change, the focus should be on those determinants which have been considered modifiable, prevalent, and relatively easy to target within children and adolescents. Again, during childhood and adolescence proved to be an important population, as behaviors which were related to obesity prevention declined during these years (Wilson, 2008).

Table 2. Modifiable behaviors related to childhood and adolescent obesity as recommended by the American Medical Association

Behaviors	
Physical Activity	Participation in at least 60 minutes of moderate to vigorous physical activity per day.
Sedentary Behaviors	Limiting screen time (television, video game, and computer use) to less than two hours per day.
Fruit and Vegetable Consumption	Eating five or more servings of fruit and vegetables daily.
Sweetened Beverage Consumption	Increasing water consumption in relation to the amount of sweetened beverages consumed.

Link between Nutrition and Health

The link between nutrition and health has been indicated time and time again. In order to achieve good health, an individual's nutritional choices become essential. Major causes of morbidity and mortality in the United States have been related to poor diet. Following a diet that complies with the Dietary Guidelines may reduce the risk of many of the chronic diseases associated with increased risk of morbidity and mortality (cardiovascular disease, hypertension, type 2 diabetes, osteoporosis, and some cancers) (USDHHS, 2005). It was estimated that about 16 % of mortality from any cause in men, and 9 % from any cause of mortality in women could be eliminated by the inclusion of recommended dietary behaviors (National Center for Education Statistics, 1996).

The link between nutrition and health for children and adolescents is of particular concern, as the relationship between their healthy growth and development has been indicated. The health of children and adolescents is dependent on obtaining food adequate in energy and nutrients to promote optimal physical, social, and cognitive growth and development (American Dietetic Association, 2006). Children and adolescents who do not meet these standards are at risk for a variety of poor outcomes, including growth retardation, iron deficiency anemia, poor academic performance, development of psychosocial difficulties, and an increased likelihood of developing the aforementioned chronic diseases. Not only has a link between delayed physical, psychosocial, and cognitive development been established, but the relationship between the growing problem of child and adolescent obesity cannot be ignored.

Benefits of a Health and Nutrition Education Curriculum

The benefits of a health and nutrition education curriculum were far reaching. Health and nutrition education have been known to help students with positive strides in their overall lifestyles, development of good personal habits, avoidance of high-risk behaviors, and with control over their own future (Merritt, 2008).

In fact, the American Dietetic Association (2005) has acknowledged the importance of nutrition education in their position statement:

It is the position of the American Dietetic Association that all children and adolescents, regardless of age, sex, socioeconomic status, racial diversity, ethnic diversity, linguistic diversity, or health status, should have access to food and nutrition programs that ensure the availability of a safe and adequate food supply that promotes optimal physical, cognitive, social, and emotional growth and development. Appropriate food and nutrition programs include food assistance and meal programs, nutrition education initiatives, and nutrition screening and assessment followed by appropriate nutrition intervention and anticipatory guidance to promote optimal nutrition status.

With the mention of both "nutrition education initiatives" and "appropriate nutrition intervention" the American Dietetic Association (2006) has recognized that all children should be provided with the knowledge and skills necessary to make healthy nutrition choices. The importance of these strategies in reducing the prevalence of obesity among children and adolescents should be a main focus now and into the future.

Importance of Schools

Traditionally, one of the primary missions of schools has been to help children grow into healthy and productive adults (Merritt, 2008). Schools were considered an important venue for delivery of health education programs since a majority of the nation's youth can be reached (Goutham, 2008). In 2000, 53.2 million students were enrolled in public and private elementary and secondary schools in the United States (IOM, 2005). Children spent a majority of their hours in school, and all children regardless of demographics had the capability of being reached during school time. In addition, nutrition and physical activity behaviors were addressed in some capacity everyday through the choice of school meals, health and/or physical education curriculum (Procter et al., 2008). Schools also had other resources which increased the viability of school-based programming such as: classrooms, educational resources, educated individuals to supervise activities/programs, and access to additional health services including school nurses (Story, 1999). Furthermore, the school is an ideal environment when dealing with peer influence and social norm expectations in relation to eating behaviors related to obesity (Procter et al., 2008).

The existing educational mission of schools was suggested to be in line with the dissemination of health information (Pyle et al., 2006). However, today's schools have faced intense pressure to focus on standardized tests, and therefore have shifted away from an emphasis on a healthy mind and a healthy body (Wechsler, McKenna, Lee, & Dietz, 2004). An increasing number of educators have realized, as the National Association of State Boards of Education (NASBE) stated, "Health and success in school are interrelated. Schools cannot achieve their primary mission of educating if students and staff are not healthy and fit physically, mentally, and socially" (NASBE, 2000, p. 2). Schools can help to meet the goal of obesity reduction, i.e. healthier students, while boosting academic achievement and maintaining or improving their own financial situation (AFHK, 2004).

Both inside and outside of the classroom, schools present opportunities for students to learn about healthful eating habits and to make food choices during school meal times and through school-related activities (AFHK, 2004; IOM, 2005). Schools have been identified as important channels through which important behavior changes related to the reduction of childhood obesity could be addressed (Carter, Wiecha, Peterson, & Gortmaker, 2002; CDC, 1997; Wilson, 2008). School-based programs have also been found to be cost-effective (Wang, Yang, Lowry, & Wechsler, 2003). Hence, the Institute of Medicine (2005) in its *Preventing Childhood Obesity: Health in the Balance* report recommended that schools should provide a consistent environment that was conducive to healthful eating behaviors. School efforts to make policy and environmental changes have not gone unnoticed, however "It is not enough to change the food on the plate. We must also provide the knowledge and the skills that enable children to make choices that lead to a nutritious diet and improved health" (National Center for Education Statistics, 1996, p.2).

Students have also been receptive to the idea of school-based initiatives. When interviews were conducted with overweight students, they responded positively, given that the program would be presented in a supportive environment, was informative, and was tailored to their needs and concerns (Neumark-Sztainer, Martin & Story, 2000). A later study revealed similar findings, with additional suggestions for the program such as low cost, convenient, entertaining, inclusion of a variety of activities, and offered to all students regardless of their weight (Neumark-Sztainer, Story, & Martin, 2003). Wilson (2008) recommended the importance of incorporating what children and adolescents viewed as important into school-based nutrition education programs.

School Curriculum

Health and nutrition curriculum are incorporated in both the National Health Education Standards and the National Science Education Standards. In addition, there are also state health standards embedded in the curriculum, with most states having their own guidelines for curriculum and instruction (Merritt, 2008). It has been approximated that more than 80 percent of school districts in the U.S. are required to teach health education. However, the effectiveness of health education curricula has been faulted, as standards are not consistent across the board.

The general priorities in health instruction have included information on diet, eating disorders, and obesity prevention. More specifically, the following topics were constituted as "priorities" in nutrition education: children's food-related habits and eating behaviors; the relationship of food, diet and nutrients to health; the links between nutrition and cardiovascular health; the food-guide pyramid; finding and choosing healthy foods and increasing positive eating patterns; nutrients and their food sources; dietary guidelines and goals that emphasize low-fat, low-salt and increased complex-carbohydrate foods. Within these topics, curriculum must be developmentally appropriate. Table 3 summarizes the main nutrition curricular concepts underlying grade-specific National Science Education Standards.

Table 3. National Science-Education Standards – Nutrition

Curricular Area: Nutrition	Statement of Curricular Concept/Principle
Grades K-4	Nutrition is essential to health. Students should understand how the body uses food and how various foods contribute to health. Recommendations for good nutrition include eating a variety of foods, eating less sugar and eating less fat.
Grades 5-8	Food provides energy and nutrients for growth and development. Nutrition requirements vary with body weight, age, sex, activity and body functioning.
Grades 9-12	Selection of foods and eating patterns determine nutritional balance. Nutritional balance has a direct effect on growth, development and personal well-being. Personal and social factors – such as habits, family income, ethnic heritage, body size, advertising and peer pressure – influence nutritional choices.

Review of Nutrition Education School-Based Obesity Prevention Programs

Lessons need to be learned from past, present and future programs which include nutrition education as a means to prevent obesity in children and adolescents. In settings where time, teacher preparation, cost, student attention span, and limited resources are constraints, measuring the effectiveness of such programs becomes imperative (Abood, Black & Coster, 2008). Several reviews have been conducted which highlight the vast array of nutrition education programming being included in school-based obesity prevention programs. See Table 4 for a summary of these interventions.

Components of Successful Programs

As several reviews have indicated, implementing nutrition education programs in the school does not necessarily ensure success. However, these reviews have also shown common components amongst successful programs (Table 5). The inclusion of all three components: the cognitive, the affective (conative) and the behavioral (volitional) has led to numerous successful nutrition education programs (Contento, Balch, & Bronner, 1995). The cognitive component focuses on providing factual nutrition information. The affective component includes beliefs, attitudes, and perceptions associated with healthy nutrition behaviors. The behavioral component supports the change process, goal-setting, skill building and readiness. Lessons have been learned within each of these components and should be considered when designing successful nutrition education obesity prevention programs in the school setting. In addition, programs complemented with physical activity were more successful in altering BMI outcomes.

Table 4. Summary of nutrition education programs

Program/ Author	Purpose	Theory	Sample	Description of Program/Nutrition Component	Major Findings	Author Suggestions
Choice, Control & Change Contento et al., 2007	To encourage personal agency and competence in both nutrition and physical activity behaviors.	TPB	Middle school students in 19 science classes within 5 schools (n=278)	24-sessions The curriculum used a unique inquiry-based scientific approach.	Students increased consumption of fruits and vegetables, decreased the frequency of sweetened beverages, packaged snacks, and eating at fast food restaurants.	Through this approach, students developed an in-depth understanding of the rationale for taking action, and the expected barriers, an increase in self-efficacy, and a sense of empowerment.
Fit For Life Gombosi et al., 2007	To investigate a primary obesity prevention program focusing on both nutrition and physical activity as a way to a healthier lifestyle, also recognizing the importance of family and community.		Students in grades K-8 and their families. (n = 4241 ± 200/year)	American Heart Association kits provided to nearly all classrooms. The Virtual Reality Wellness Club, a series of work booklets that progressively challenge participants to track their physical activity and nutrition choices.	Outcome data did not show significant improvement in knowledge or BMI outcomes.	Helped enhance the use of a standard health curriculum.
Frenn's Study Frenn et al., 2003	To reduce fat intake in diet and increase physical activity in middle school students.	Health Promotion Model, TM	$6^{th} - 8^{th}$ grade, experimental group (n = 60) and control (n = 57)	4 classroom 45-min. sessions based on stage of change Focus on consciousness raising and self-reevaluation. Use of age-specific examples, games, preparation and sampling of snacks.	Percentage fat in food was sig. less and duration of exercise for intervention group was sig. higher.	Short interventions might be viable options in the right setting.
Louisiana Health Williamson et al., 2008	To modify environmental cues, enhance social support, and promote self-efficacy for health behavior change.	SCT	17 school clusters assigned to one of three intervention arms (n = 775, 727, 600)	Program delivered as part of regular classroom instruction, combined with internet counseling and email communications for children and their parents. Children will receive weekly lessons on healthy eating and exercise that include a 20–25 min interactive didactic segment followed by hands-on activities.	Results to come.	The target population represent poor families who could not readily access appropriate health care or preventive services.

Table 4. (continued)

Program/ Author	Purpose	Theory	Sample	Description of Program/Nutrition Component	Major Findings	Author Suggestions
Michigan Model Nutrition Curriculum Fahlman et al., 2008	To determine the impact of the curriculum on nutrition knowledge, efficacy expectations, and eating behaviors.		Middle school students divided into an intervention group (n = 407) and a control group (n = 169).	Teachers were trained to incorporate the 8 lessons into their curriculum. Topics related to nutrition knowledge such as the food groups, food pyramid, food labels, advertising, and body image. They also contained components specifically designed to target nutritional risk behaviors.	In the intervention group: increased nutrition knowledge, significantly more likely to eat FV and less likely to eat junk food, and felt more confident that they could eat healthy.	Well-designed and -executed school-based nutrition programs can result in positive changes in dietary eating behaviors, as well as increased self-efficacy expectancy regarding nutritional choices in middle school children. Further research needed to determine the long-term impact.
New Moves Neumark-Stainer et al., 2003 Flattum et al., 2009	To increase physical activity, healthful eating, and social support when implemented as a multi-component, girls-only PE class.	SCT	14-17 yr. old inactive adolescent high school girls (n = 41)	16-week program (once every other week) Focused on skill building and self-efficacy through use of small groups, class discussions, role play, preparing and eating healthy snacks. Follow-up sessions included individual sessions, using motivational interviewing to develop goals and actions related to eating and physical activity.	Majority of initial outcomes were not significant. Among the 20 girls who were a part of the motivational interviewing, 81% completed all of the individual sessions, and girls set a goal 100% of the time.	Gender specific education may help those who face barriers in traditional co-ed classes. Motivational interviewing offers a promising component of school-based obesity-prevention programs and was found to be feasible to implement in school settings and acceptable to the adolescents.
Nutrition in the Garden McAleese & Rankin, 2007	To investigate the effects of garden-based nutrition education on adolescents' fruit and vegetable consumption.		Sixth grade students (n=99) at three different elementary schools made up a control and two treatment groups.	The 12-week curriculum provided lessons and activities that combined nutrition and hands-on, garden-based activities. Students were also involved in food experiences with FV harvested from the garden.	Adolescents who participated in the garden-based nutrition intervention increased their servings of FV more than students in the two other groups.	These results help to show the importance of hands-on activities when attempting to change nutrition-related behavior such as fruit and vegetable consumption.

Table 4. (continued)

Program/ Author	Purpose	Theory	Sample	Description of Program/Nutrition Component	Major Findings	Author Suggestions
Pathways Caballero et al., 2003	To evaluate the effectiveness of a school-based multi-component intervention for reducing body fat in American Indian children including classroom curriculum, food service modifications, physical activity and family involvement.	SCT	3rd – 5th grade (n = 1,704)	12-weeks (2 x 45-min. lessons) Classroom curriculum was based on relevant culture and practices.	Knowledge, attitudes and behaviors were positively and sig. changed.	Curricular changes can improve knowledge and behaviors. Longer interventions may be needed to see sig. changes in percent body fat.
Pilot Community Prevention Program Hawley et al., 2008	To educate students and families about the importance of good nutrition and exercise and to motivate them to reach these goals by using principles of behavior change.	TM	Sixth graders (n = 65) and their families	The program consisted of five 40-min sessions during physical education classes, over the course of 6 weeks. Instruction was in nutrition, fitness, goal setting, and self-efficacy. Sessions consisted of adventure-based games and tasks. A family component helped get students started and focused on setting and working towards goals.	There were no sig changes in students' individual health attitudes and behaviors. However, sig changes did occur among families -the goal of eating healthfully was sig more important.	The preliminary study provided a strong foundation on which to build an intervention that addressed the community's particular nutrition and activity challenges.
Planet Health Gortmaker et al., 1999	To integrate the health goals of increasing physical activity, decreasing TV viewing, improving FV intake, and moderating fat intake into the middle school curriculum.	SCT	6th – 8th grade, implemented with 129 teachers in 6 schools	32 classroom lessons integrated into other disciplines	Reduced obesity in girls, increased FV consumption, reduced TV watching.	Found to be feasible and acceptable when implemented in the school setting.

Table 4. (continued)

Program/ Author	Purpose	Theory	Sample	Description of Program/ Nutrition Component	Major Findings	Author Suggestions
Policy-based school intervention Foster et al., 2008	To examine the effects of a multi-component, school nutrition policy initiative; and show how food choices and physical activity are tied to personal behavior, individual health, and the environment.		Participants were 1349 students in 4th – 6th grade from 10 schools with 50% of students eligible for free or reduced-price meals.	The goal was to integrate nutrition (50 hours/year) into various disciplines. The educational component was designed to be integrative and interdisciplinary.	There was a 50% reduction in the incidence of overweight. Significantly fewer children in the intervention schools (7.5%) than in the control schools (14.9%) became overweight after 2 years.	A multicomponent school-based intervention can be effective in preventing the development of overweight.
Present and Prevent Abood et al., 2008	To assess a school-based nutrition minimal intervention which used a commercially available power-point program.		Middle school students (mean age = 14.5), experimental group (n = 551), delayed treatment group (n = 329)	Program presented in two 30-minute time slots over 1 week. Included information on : health problems associated with and causes of obesity, importance of preventing obesity, body image, benefits of and achieving healthy body weight, healthful food choices, reading food labels, portion control, how to change unhealthful habits, how to start being more active and overcoming barriers.	Significant experimental improvement occurred in: knowledge, intention to maintain a healthy body weight because of importance to friends, intention to eat fewer fried foods, eat fewer sweets, look more at food labels, and limit TV watching.	The impact of the minimal program should be considered with decreased school resources, as well as the attention span and interest level of teens. Nutrition education MIs may serve as an adjunct to existing approaches to obesity prevention
USDA fresh fruit and vegetable program Jamelske et al., 2008	Whether an impact was made by the FFVP, which provided funding for students to receive a free fresh fruit or vegetable snack daily for an academic year.		4th, 7th, and 9th graders (n = 1127) 784 students in 10 intervention schools and 343 students in 10 control schools	Students were provided with fruit and vegetable snacks daily for a year to see if exposure to different fruits and vegetables would alter attitudes, behaviors.	Intervention students reported an increased willingness to try new fruits (24.8% versus 12.8%) and vegetables (25.1% versus 18.4%) at school.	Consumption of FV as a habit in childhood is an important predictor of higher consumption as adults. It is essential that the effects of school, district, or state policy changes regarding the school food environment are evaluated.

Table 4. (continued)

Program/ Author	Purpose	Theory	Sample	Description of Program/ Nutrition Component	Major Findings	Author Suggestions
Ward-Begnoche et al., 2008	To incorporate nutritional, physical activity, and coping skills curriculum that emphasized education and behavioral change into the PE curriculum.		Middle school students (6th – 8th grade) taking physical education (n = 78)	16-sessions including worksheets, handouts, and structured activities. Small group format was used to aid in rapport building with instructors. Information included food labels, review of popular fast food restaurants' nutritional information, and portion size.	Results to come.	Benefits identified included helping the school to establish a nutrition and physical activity goal for their school, provide teachers and personnel an opportunity to facilitate goal-setting and behavior-change curriculum.

Nutrition education programs which promoted healthy weight rather than focusing on weight loss or attaining a specific weight were less likely to intensify diet and self-reported insecurities (Pyle et al., 2006). It was recommended for programs to focus on improving nutrition habits in combination with physical activity, and other healthy lifestyle choices.

Table 5. Components of successful nutrition education programs

Behavioral focus
Theoretical framework
Adequate time and intensity
Family involvement
Multicomponent strategies
Developmentally appropriate
Considers needs of students, teachers and school
Strengthen self-efficacy, skills
Culturally relevant
Evaluation

An essential component of these nutrition education programs, either primary or secondary, was a focus on behavior change (Caballero, 2004). Most successful programs combined several different techniques including: behavior management counseling, teaching strategies to identify healthy foods and avoid unhealthy foods, implementing healthy nutritional choices into the daily menu, and involving parents and other family members throughout the program (Pyle et al., 2006). To be effective, programs should focus on promoting healthy behaviors in a way to impact consequent behavior into and throughout adulthood.

Nutrition education was an integral element within successful child and adolescent school-based obesity prevention programs. Providing age-appropriate, culturally sensitive instruction and materials was connected to the positive behavioral outcomes of the programs (AFHK, 2004). Effective programs in diverse communities must be especially cognicent of individuals and their backgrounds, and tailor the programs as such (Perez-Rodrigo & Aranceta, 2003). Casazza and Ciccazzo (2006) indicated the effectiveness of a nutrition-education program when relevant to the student, as well as the applicability of the information to the student's everyday life. Focusing on learning styles of "today's" children and adolescents becomes vital. Computer-based, interactive, animated education programs were effective in eliciting nutrition behavior changes.

To adopt, maintain, and enjoy healthy nutrition habits it is critical to go beyond dissemination of knowledge and to promote the development of attitudes, skills, and behaviors. Use of behavioral theories often help in going beyond simple dissemination of knowledge. Examples of behavioral theories are: health belief model, social cognitive theory, transtheoretical model, theory of reasoned action, and theory of planned behavior. McAleese, Rankin and Fada (2007) established the importance of hands-on activities, such as schoolyard gardens, when attempting to change nutrition-related behaviors. McAleese et al. (2007) found children and adolescent to be more receptive when they were involved with a project related to their desired nutrition goals.

However, in such programs nutrition education also went further than the expected classroom education. Schools have been described as having the opportunity to influence the nutrition children receive, as well as helping to establish healthy lifelong habits by including school personnel in modeling healthy choices; advocating for healthy eating by providing healthy school lunches and snacks; exposing students to the selection, preparation, and taste of healthful foods (Pyle et al., 2006). Since nearly every school in the nation served at least one meal a day, five days a week, the impact of these changes, coupled with effective nutrition education could be immense (Haskins, Paxson, & Donahue, 2006).

Challenges of Nutrition Education in School-Based Programs

Although schools seem to be an ideal setting to implement nutrition education programs which could contribute to obesity prevention, there are challenges which present roadblocks to implementation. Increasing curricular mandates, time demands on teachers, and financial strain on the school systems have all contributed to the decreased importance placed on health education, including nutrition education, within the school curriculum (Merritt, 2008). Many times these subjects are overlooked or taught briefly within other subjects, thereby decreasing the effectiveness of learning.

The ever-changing diverse student body also presents a challenge to the implementation of nutrition education programs (Merritt, 2008). Many students come from very different backgrounds, including differences in race/ethnicity, cultural beliefs, socioeconomic status, and parental involvement just to name a few. With recommendations hinging on the information presented being relevant to the student, this creates some challenges.

Parental buy-in plays a role in the efficacy of nutrition education programs. Unfortunately most parents do not set good examples for their children. Without the inclusion and responsiveness of parents, the educational messages taught throughout the nutrition program may be lost (Merritt, 2008).

Adequate training to teach the nutrition education programs has often been overlooked. In order for teachers to teach the curricula effectively, they must be properly trained (Merritt, 2008). This typically involves time and money which the school systems do not have. Having a support person for such a program also tends to help the implementation, as well as the sustainability of the program. Most schools do not have funds allocated for such a position, leaving the coordination of said program up in the air.

Recommendations for School-Based Nutrition Education Interventions

With such a huge focus on childhood obesity, numerous leaders in health education have proposed recommendations to improve the success of school-based programs. The Institute of Medicine (2005) realized that teaching children the importance of health promotion practices was a priority and school was considered an ideal place for promoting habits that would support a healthy future. The IOM suggested schools must take the opportunity to provide lessons in healthy behaviors that will allow children to be more effective learners, as well as provide a consistent environment conducive to healthful eating behaviors. In order for this to happen, schools should be provided with adequate and sustained federal and state government funding to implement changes in the school environment and provide appropriate nutrition education (IOM, 2006).

School environmental factors included improved school health curricula (with innovative approaches to teaching and staffing); school health services for obesity prevention efforts; school meals that met the Dietary Guidelines for Americans; nutritional standards for all competitive foods and beverages sold or served in schools (IOM, 2006). In addition, it was recommended that schools conduct an annual assessment of students' weight, height, and body mass index and make the information available to parents in an effort to identify at-risk children.

In addition to the school environment, personal, behavioral, social and community environmental factors must be considered. School-based nutrition education programs should consider the needs and interests of students, teachers, and the school (Perez-Rodrigo & Aranceta, 2003). As reported in a review by Doak et al. (2006), effective programs need to be tailored to the individual needs of the children, according to gender and ethnicity. Future interventions should also address the psychological and environmental influences of the home environment through education and active involvement of parents. Developing partnerships with parents and communities would allow schools to fight childhood obesity more effectively and ensure a healthy future for the children (IOM, 2005; Rosenthal & Chang, 2004).

Beyond the content of the nutrition education programs, it was noted that obesity prevention messages should be framed positively. Focusing on potential solutions rather than on problems was more likely to cause people to take action. The relationship between children's emotional well-being in relation to overweight and obesity was not assessed in most programs. Doak et al. (2006) warned that singling out children at risk may contribute to their emotional and mental distress. The impact of the children's emotional well-being indicated the need to further explore not only the specific messages being targeted, but also the delivery and implementation of the program.

CDC Recommendations. The Centers for Disease Control and Prevention (CDC, 1997) published guidelines that identified school policies and practices most likely to be effective in promoting lifelong physical activity and healthy eating. The guidelines were directly related to many of the recommendations for school-based interventions. As such, the CDC recommendations should be considered important when initiating nutrition-related changes in school environments and programming. The guidelines applicable to school-based nutrition education programs are as follows: 1) address nutrition through a Coordinated School Health Program (CSHP) approach; 2) designate a school health coordinator and maintain an active school health council; 3) assess the school's health policies and programs and develop a plan for improvement; 4) strengthen the school's nutrition policies; 5) implement a high-quality health promotion program for school staff; 6) implement a high-quality course of study in health education; 7) implement a quality school meals program; and 8) ensure that students have appealing, healthy choices in foods and beverages offered outside of the school meals program.

Research has shown that in schools where nutrition education efforts are coordinated, a more focused message about the importance of healthy eating is presented. Integration across the curriculum and building on lessons from year to year can only enhance the success of nutrition education programs in the school setting (National Center for Education Statistics, 1996).

Conclusion

The need for improved nutrition among America's youth is evident, as the rates of child and adolescent overweight and obesity continue to increase. The complexity of the issue has made a solution difficult. It is understood that schools cannot solve the obesity epidemic on their own (Wechsler et al., 2000), but school-based nutrition education programs have the potential for reaching numerous children in a cost-effective manner (Plye et al., 2006). The former U.S. Surgeon General has been quoted as saying, "I do not blame schools for our obesity epidemic. Instead I look to schools – and everyone who has an influential hand in education – as a powerful force for change" (Office of the Surgeon General, 2001). Skills taught in school-based programs are accepted as the foundations for healthy lifestyles, regardless of children's weight status. As expressed by those involved with Action for Healthy Kids, providing the resources and tools which evoke change was a necessity. Just as the problems which have led to obesity were multifaceted, the solutions need to be as well (AFHK, 2004).

Development of healthy lifestyle habits that persist into adulthood, and thus, have a long-term impact on the health and well-being of future generations is critical (USDHHS, 2007). The diversity of nutrition education programs which have been effective has been promising: the likelihood of finding a suitable program, adaptable to specific community needs is upon us (Doak et al., 2006). Nevertheless, nutrition education programs must be developed around community needs, and guided by those which have already been successful. As Doak et al. stated, "There is no need to reinvent the wheel" (p.129). Building on successful programs and relating them to the child, adolescent, and specific community needs will increase the likelihood of success. These significant lessons must be used to make an impact on the childhood obesity epidemic which continues to plague our nation.

References

Abood, D., Black, D. & Coster, D. (2008). Evaluation of a school-based teen obesity prevention minimal intervention. *Journal of Nutrition Education and Behavior, 40,* 168-174.

Action for Healthy Kids. (2004). *The value of improving nutrition and physical activity in our schools.* Retrieved July 9, 2008 from www.actionforhealthykids.org

American Dietetic Association. (2006). Position of the American Dietetic Association: Child and adolescent food and nutrition programs. *Journal of the American Dietetic Association, 106,* 1467-1475.

Caballero, B., Clay, T., Davis, S., Ethelbath, B., Rock, B., Lohman, T., et al. (2003). Pathways: A school-based randomized controlled trial for the prevention of obesity in American Indian school children. *American Journal of Clinical Nutrition, 78,* 1030-1038.

Caballero, B. (2004). Obesity prevention in children: Opportunities and challenges. *Journal of Obesity, 28,* S90-S95.

Carter, J., Wiecha, J., Peterson, K. & Gormaker, S. (2001). *An interdisciplinary curriculum for teaching middle school nutrition and physical activity.* Champaign, IL: Human Kinetics.

Casazza, K., & Ciccazzo, M. (2006). Improving the dietary patterns of adolescents using a computer-based approach. *Journal of School Health, 76,* 43-46.

Centers for Disease Control and Prevention. (1997). Guidelines for school and community programs to promote lifelong physical activity among young people. *MMWR, 46,* 1-36.

Centers for Disease Control and Prevention, National Center for Health Statistics. (2005). *National health and nutrition examination survey (NHANES).* Retrieved October 10,2007 from http: www.cdc.gov/nchs/about/major/nhanes/nhanes2005-2006/nhanes05_06.html

Contento, I., Balch, G., & Bronner, Y. (1995). Theoretical frameworks of models for nutrition education. *Journal of Nutrition Education, 27,* 287-290.

Contento, I., Koch, P., Lee, H., Sauberli, W. & Calabrese-Barton, A. (2007). Enhancing personal agency and competence in eating and moving: Formative evaluations of a middle school curriculum – Choice, control, and change. *Journal of Nutrition Education Behavior, 39,* 179-186.

Cooley, C. (1964). *Human nature and the social order.* New York: Schocken.

Cosrow, N., Jeffrey, R., & McGuire, M. (2001). Understanding weight stigmatization: A focus group study. *Journal of Nutrition Education, 33,* 208-214.

Doak, C. M., Visscher, T., Renders, M. & Seidell, J. C. (2006). The prevention of overweight and obesity in children and adolescents: A review of interventions and programmes. *Obesity Reviews, 7,* 111-136.

Erickson, S., Robinson, T., Haydel, K. & Killen, J. (2000). Are overweight children unhappy. *Archives of Pediatrics and Adolescent Medicine, 154,* 931-934.

Evans, W., Finkelstein, E., Kamerow, D., & Renaud, J. (2005). Public perceptions of childhood obesity. *American Journal of Preventive Medicine, 28,* 26-32.

Fahlman, M., Dake, J., McCaughtry, N., & Martin, J. (2008). A pilot study to examine the effects of a nutrition intervention on nutrition knowledge, behaviors, and efficacy expectations in middle school children. *Journal of School Health, 78,* 216-222.

Finkelstein, E., Fiebelkorn, I., Wang, G. (2003). National medical spending attributable to overweight and obesity: How much, and who's paying? *Health Affairs, 3,* 219–226.

Flattum, C., Friend, S., Neumark-Sztainer, D., & Story, M. (2009). Motivational interviewing as a component of a school-based obesity prevention program for adolescent girls. *Journal of American Dietetic Association, 109,* 91-94.

Foster, G., Sherman, S., Borradalie, K., Grudny, K., Vander Veur, S., Nahcmani, J., et al. (2008). A policy-based school intervention to prevent overweight and obesity. *Pediatrics, 121,* 794-802.

Freedman, D. S., Dietz, W. H., Srinivasan, S. R. & Berenson, G. S. (1999). The relation of overweight to cardiovascular risk factors among children and adolescents: *The Bogalusa*

Frenn, M., Malin, S. & Bansal, N. (2003). Stage-based intervention for low-fat diet with middle school students. *Journal of Pediatric Nursing, 18,* 36-45.

Gombosi, R., Olasin, R. & Bittle, J. (2007). Tioga county fit for life: A primary obesity prevention project. *Clinical Pediatrics, 46,* 592-600.

Gortmaker, S., Peterson, Wiecha, J., Sobaol, A., Dixit, S., Fox, M. & Laird, N. (1999). Reducing obesity via school-based interdisciplinary intervention among youth: Planet Health. *Archives of Pediatric Adolescent Medicine, 153,* 409-418.

Goutham, R. (2008). Childhood obesity: Highlights of AMA expert committee recommendations. *American Family Physician, 78,* 56-64.

Haskins, R., Paxson, C., & Donahue, E. (2006). *Future of children: Fighting obesity in the public schools.* Princeton Brookings. Retrieved July 9, 2008 from www.futreofchildren.org

Hawley, S., Beckman, H. & Bishop, T. (2006). Development of an obesity prevention and management program for children and adolescents in a rural setting. *Journal of Community Health Nursing, 23,* 69-80.

Heart Study. *Pediatrics, 103,* 1175-1182.

Institute of Medicine. (2004). *Preventing childhood obesity: Health in the balance.* Washington, DC: National Academy Press.

Institute of Medicine. (2005). *Preventing Childhood Obesity: Health in the Balance.* Retrieved July 9, 2008 from www.iom.edu

Institute of Medicine (2006). *Progress in Preventing Childhood Obesity: How Do We Measure Up?* Washington, DC: National Academies Press, Retrieved July 9, 2008 from www.nap.edu

Jamelske, E., Bica, L., McCarty, D. & Meinen, A. (2008). Preliminary findings from an evaluation of the USDA fresh fruit and vegetable program in Wisconsin schools. *Wisconsin Medical Journal, 107,* 225-230.

Kaufman, F. R. (2003). Type 2 diabetes in children and youth. *Reviews in Endocrine & Metabolic Disorders, 4,* 33-42.

Lobstein, T., Baur, L., Uauy, R. & International Association for the Study of Obesity Task Force. (2004). Obesity in children and young people: A crisis in public health. *Obesity Reviews, 5*(Suppl. 1), 4-104.

McAleese, J. & Rankin, L. (2007). Garden-based nutrition education affects fruit and vegetable consumption in sixth-grade adolescents. *Journal of the American Dietetic Association, 107,* 662-665.

Merritt, R. D. (2008). Health and nutrition studies: Curriculum organization – Health and nutrition in the curriculum. Research Starters: EBSCO Publishing Inc.

Moreno, L. A., Tomas, C., Gonzalez-Gross, M., Bueno, G., Perez-Gonzalez, J. M. & Bueno, M. (2004). Micro-environmental and socio-demographic determinants of childhood obesity. *International Journal of Obesity and Related Metabolic Disorders, 28* (Suppl. 3), S16-S20.

National Association of State Boards of Education. (2000). *Fit, healthy, and ready to learn.* Alexandria, VA: Author.

National Center for Education Statistics. (1996). Nutrition education in public elementary and secondary schools.

Neumark-Sztainer, D., Martin, S.L. & Story, M. (2000). School-based programs for obesity prevention: What do adolescents recommend? *American Journal of Health Promotion, 14,* 232–235.

Neumark-Sztainer, D., Story, M. & Martin, S. (2003). New moves: A school-based obesity prevention program for adolescent girls. *Preventive Medicine, 37,* 41-51.

Office of the Surgeon General. (2001). *The Surgeon General's call to action to prevent and decrease overweight and obesity.* Rockville, MD: Author.

Ogden, C. L., Flegal, K. M., Carroll, M. D. & Johnson, C. L. (2002). Prevalence and trends in overweight among U.S. children and adolescents, 1999-2000. *Journal of the American Medical Association, 288,* 1728-1732.

Perez-Rodrigo, C. & Aranceta, J. (2003). Nutrition education in schools: Experiences and challenges. *European Journal of Clinical Nutrition, 57,* S82-S85.

Procter, K. L., Rudolf, M. C., Feltbower, R. G., Levine, R., Connor, A., Robinson, M. et al. (2008). Measuring the school impact on child obesity. *Social Science and Medicine, 67,* 341-349.

Pyle, S., Sharkey, J., Yetter, G., Felix, E. & Furlong, M. (2006). Fighting an epidemic: The role of schools in reducing childhood obesity. *Psychology in the School, 43,* 361-375.

Rosenthal, J. & Chang, D. (2004). *State approaches to childhood obesity: A snapshot of promising practices and lessons learned.* National Academy for State Health Policy: Portland, ME.

Schwartz, M. & Puhl, R. (2003). Childhood obesity: A societal problem to solve. *Obesity Reviews, 4,* 57-71.

Schwimmer, J., Burwinkle, T. & Varni, J. (2003). Health-related quality of life of severely obese children and adolescents. *The Journal of the American Medical Association, 289,* 1813-1819.

Shelton. (2008). Physicians increasingly seeing overweight children with "adult" diseases. *Chicago Tribune, 3/19/08.*

Smith, J. (2004). The current epidemic of childhood obesity and its implications for future coronary heart disease. *Pediatric Clinics of North America, 51,* 1679-1695.

Story, M. (1999). School-based approaches for preventing and treating obesity. *International Journal of Obesity, 23,* S43-S51.

Strauss, R. S. (2000). Childhood obesity and self esteem. *Pediatrics, 105,* E15.

U.S. Department of Health and Human Services. (2004). *Prevalence of overweight among children and adolescents: United States 2003-2004*. Atlanta, GA:

U.S. Department of Health and Human Services, Centers for Disease Control and Prevention. Retrieved March 1, 2009 from
http://www.cdc.gov/nchs/products/pubs/pubd/hestats/overweight/overwght_child_03.htm

U.S. Department of Health and Human Services & U.S. Department of Agriculture. (2005). *Dietary Guidelines for Americans*, 2005 (6th Edition). Washington, DC: U.S. Government Printing Office.

U.S. Department of Health and Human Services. (2007). *Overweight and obesity*. U.S. Department of Health and Human Services, Centers for Disease Control and Prevention. Retrieved March 1, 2009 from
http://www.cdc.gov/nccdphp/dnpa/obesity/childhood/index.htm

Wang, L., Yang, Q., Lowry, R. & Wechsler, H. (2003). Economic analysis of a school-based obesity prevention program. *Obesity Research, 11,* 1313-1324.

Ward-Begnoche, W., Gance-Cleveland, B., Harris, M. & Dean, J. (2008). Description of the design and implementation of a school-based obesity prevention program addressing needs of middle school students. *Journal of Applied Clinical Psychology, 24,* 247-263.

Wechsler, H., McKenna, M., Lee, S. & Dietz, W. (2004). The role of schools in preventing childhood obesity. *The State Education Standard, December, 4-12.*

Whitaker, R., Wright, J., Pepe, M., Seidel, K. & Dietz, W. (1997). Predicting obesity in young adulthood from childhood and parental obesity. *New England Journal of Medicine, 337,* 467-474.

Williamson, D., Champagne, C., Harsha, D., Han, H., Martin, C., Newton, R., et al. (2008). Louisiana (LA) health: Design and methods for a childhood obesity prevention program in rural schools. *Contemporary Clinical Trials, 29,* 783-795.

Wilson, L. (2008). Adolescents' attitudes about obesity and what they want in obesity prevention programs. *The Journal of School Nursing, 23,* 229-238.

World Health Organization (2003). Obesity and overweight, 2003. Retrieved March 1, 2009 from *http://www.who.int/dietphysicalactivity/publications/facts/obesity/en/prin*

In: Childhood Obesity: Risk Factors, Health Effects...   ISBN: 978-1-61761-982-3
Editor: Carol M. Segel   ©2011 Nova Science Publishers, Inc.

Chapter XII

Child Obesity in Global Perspective Emergent Risks Related to Social Status, Urbanism, and Poverty

Alexandra Brewis[a] and Mary C. Meyer[b**]*
[a] School of Human Evolution and Social Change,
Arizona State University, Tempe, AZ, USA
[b] Department of Statistics, University of Georgia, Athens, GA, USA

Abstract

Very little is known currently about the pattern of risk for early childhood overweight and obesity in the least developed countries, where child under-nutrition remains very common and a pressing concern. We use standardized anthropometric and interview data pertaining to seventeen nationally-representative samples of 37,714 children aged between 30 and 60 months to model that risk. We particularly consider the possible roles of changing social and economic status of households and urban residence, and take into account such factors as variations in family size, in maternal nutritional status, and children's histories of under-nutrition (observed as growth stunting). The relationships among these variables are quite different across world regions except for mothers' overweight status, which was a strong predictor in all. In sub-Saharan Africa, overweight children are extremely rare and the only strong predictor is having a mother who is overweight. In Northern Africa urban residence is a risk factor. In the Americas, increasing wealth and social status of households raises risk substantially. Stunting places children in Africa, but not the Americas, at significantly increased risk of being overweight, and in northern Africa this effect is particularly pronounced in cities. We find every indication in these trends that child obesity and overweight might very quickly emerge as the modal nutritional status of children worldwide. The model suggests that childhood overweight in many ways embodies relative poverty as national wealth rises,

---
[*] E-mail address: alex.brewis@asu.
[**] E-mail address: mmeyer@stat.uga.edu

just as child stunting reflects the conditions of absolute poverty. As economic growth accelerates in the poorer countries, the least advantaged sectors of their populations can remain absolutely poor while their relative poverty also increases. This eans that risk for childhood obesity will grow, and probably rapidly, and it will increasingly co-exist with and so be intensified by under-nutrition.

Introduction

Child overweight and obesity have been growing rapidly in the U.S. over the last several decades, and the shifts from normal to overweight are occurring at increasingly young ages (Ogden et al. 1997). According to analyses of National Health and Nutrition Examination Survey data (NHANES), by 2002 some 23 percent of children aged 2-5 years were overweight or obese (Hedley et al. 2004). The general patterns and proximal causes of these increases of early childhood overweight and obesity in the U.S. and other wealthy nations has been both a subject of mounting policy concern and increasing research attention. We know considerably less about childhood overweight and obesity (hereafter 'overweight') beyond the industrialized world: perhaps the least is known about preschool-age children in developing countries (Martorell et al. 2000) who are our focus here. The lack of research might be explained in part by the understanding that the rates of child overweight and obesity in the developing world remain comparatively low and are seemingly stable (e.g., Monteiro et al., 2004a, 2004b, Martorell 2000:966). The underlying presumption of our chapter is that this is likely to change, and soon. Especially in middle-income developing countries, and most particularly among women, adult obesity is rising extremely rapidly, at earlier levels of economic development, and a pace much greater than that observed historically in industrialized countries (Popkin 2004). If mothers are overweight, then some of the causal mechanisms that proximally underwrite child overweight risk are likely already present.

Drawing any global picture of how and why risk of early childhood obesity is distributed in the poorer developing world has also been difficult because of methodological inconsistencies in the conduct and reporting of national studies (Shetty 1999, WHO 1998, Wang 2001). Here we use very large and comparable datasets from seventeen developing countries to model where and how young children are most at risk of being overweight. We are well able to model regional and global patterns because our samples are large enough to capture a sufficient number of cases (N=1,455 overweight children, representing 3.4 percent of the total sample) and because the individual country samples we use were all collected using standardized sampling, measurement, and interview procedures, and designed to be nationally representative. The countries range from some of the poorest to lower middle income nations, and in all the risk of being underweight greatly exceeds the risk of being overweight in early childhood (see Table 1). The rates of early childhood overweight vary from less than one percent in Haiti and Ethiopia to a high of nine percent in Egypt.

In developing the model we focus on how early childhood overweight risk is patterned across countries in relation to three core proximate factors: (1) urban-rural residence, (2) the social status of households, and (3) the economic status of households. We also consider (4) whether there might be regional effects, particularly how the relationships among these other variables might be distinct in different world regions. One of the main questions we ask here is: Is the risk of child obesity in developing countries predominantly an urban phenomenon, and if so, is this mainly explained by differences in socioeconomic status of households in urban settings compared to rural? Urbanism, social status (education and occupation), and household wealth all may certainly interact in vitally important ways to explain children's obesity risk, and need to be disassociated statistically if we are to unravel the relative influence of each. For example, urbanism and household assets are highly positively associated variables, and the relationship between poverty and overweight risk may be more relative than absolute, a product of the interaction between household and broader economic conditions. The use of a very large sample combined with careful choice of the right model allows us to distinguish the effects of these, while also taking into account a variety of other potentially confounding micro-level, familial, and life history factors, such as child's gender, growth stunting of the child, breastfeeding history, number of other children in the household, and the weight status of their mothers.

The Possible Roles of Urbanism and Socio-economics in Early Childhood Overweight in Developing Countries

What do we currently know about the likely relationship between child obesity and the related processes of urbanism and socio-economic shifts in developing countries? While some localized studies in low and middle income countries have shown a positive association between urban residence and childhood obesity risk, across-country studies have not identified any clear or consistent pattern (Wang 2001). The findings are clearer for adult women, who tend to obesity in urban areas; certainly national women's obesity rates are highest in the developing countries with the highest proportion of their population in urban centers (Mendez and Popkin 2004). Children's *under*-nutritional risk is lower in cities in the developing world, and it is suggested that this is mainly a product of an overall improvement in socioeconomic conditions in urban areas relative to rural (Smith et al. 2004). While studies with U.S. samples have concluded that once you take into account such factors as ethnicity, income, and education, the urban-rural differences in adult and child obesity shrink considerably, it is unclear at this time if this is also the case for developing countries because conclusive studies are absent. It is, however, possible that - even careful control for such aspects of social and economic status (SES) - that the distinctions between the rural and urban lifeways in developing countries might be much more differentiated than they are in the U.S. (with very different food systems, food economies, and food availability, as well as educational, occupational, and cash economy constraints and opportunities (Brody 2002)) and thus might create quite different geographies of risk. Here we test both possibilities.

SES and adult obesity is conventionally understood to have a negative relationship among adults in developed countries (Ball and Crawford 2005). Children in lower SES groups in the U.S. are at increased risk of being overweight (e.g., Crooks 1995), just as is the case for adults in the U.S. and other high income countries such as Britain, Australia, and in Europe (Jebb et al. 2003, Livingstone 2000, Margery et al. 2001). The relationship between SES and obesity in developing countries, by contrast, is considered inverted, whereby the greatest risk of being overweight in poorer countries is among the relatively wealthy. This view derives substantively from a review of the studies in developing countries conducted prior to 1989 showing that in most cases higher SES was associated with greater risk of obesity (Sobal and Stunkard 1989). Notably, although the association was quite clear and robust for women, it was less consistent for men and children. Subsequently, studies have shown the advantaged sectors of developing countries can have child overweight rates that are extremely high (e.g., Brewis 2003). And, among the very poor living in subsistence conditions, the realities of food shortages and manual labor continue to make it difficult to exceed energy balance chronically in such a way as to create any real risk of being overweight (Caballero 2005).

Studies conducted at the national level have shown that national income has found to be a strong predictor of adult obesity risk. For example, in very low GNP countries low SES has been found highly protective against adult obesity (Monteiro et al 2004a, b). But, in the middle income developing countries, where wealth is increasing rapidly, the pattern appears to be shifting. Recently the burden of adult (especially women's) obesity in middle GNP countries (defined as those with above US $2500) appears to have moved into the shifted to lower SES groups, a process that appears to be accelerating as the wealth of countries increases (Popkin 2004, Monteiro et al 2004a, b).

It is important to recognize that different measures of SES, such as household assets versus educational level, can differently predict obesity risk (Ball and Crawford 2005). Many studies that consider SES-obesity relationships in developing countries use singular measures that clump social and economic variation into a single dimension, making it more difficult to interpret findings. In our analyses, we are careful to distinguish as much as possible different aspects of the social and economic status of children proximate (household) and national circumstances, allowing improved understanding of how these might influence risk. This includes a need to differentiate how absolute and relative social and economic standing relate to risk of childhood obesity. Here we model these in relation to each other across and within regional and country contexts. By absolute wealth, we mean empirical economic and social assets (such as if you have running water, electricity, permanent flooring, a television, or car, or your educational and occupational standing). By relative wealth, we mean the relativity of the household's absolute wealth to that of others in the some country and region. That is, the absolute wealth (material assets or educational levels) in a household in the poorest sector of a higher income country could match that of a household in the wealthiest sector of a very poor country, but the relative wealth of each is completely different (low in the former versus high in the latter). Based on what has been observed in adult women's patterns in developing countries, we could predict that child overweight would be linked to relative rather than absolute wealth, That is, risk of child obesity might not necessarily increase with household wealth per se., but rather as GNP rises the greater the effect of relatively lower SES should be on young children's risk of being overweight.

Table 1: Sample sizes, survey dates, and some descriptive statistics for the seventeen individual country samples.

Country	GNI in 2002	Survey Dates	Sample Size[a]	Percent Urban[b]	Percent homes with television[c]	Percent children overweight[d]	Percent children stunted[e]	Percent children underweight[e]	Percent mothers overweight[e]	Breastfeeding duration[b,f]	Percentage of mothers with no formal education[b]
NORTH AFRICA											
Egypt	1470	2000	3998	40.7	85.6	9.0	18.7	4	70.4	3.6	43.2
Morocco	1170	2003-2004	2236	43.6	56.3	4.9	18.2	10.2	44.4	2.0	50
SUB-SAHARAN AFRICA											
Benin	380	2001	1239	32.7	17.3	1.0	30.4	22.8	19.1	2.9	64.1
Burkina Faso	250	2003	3195	16.8	9.5	2.2	38.6	37.6	7.6	7.2	80.3
Ethiopia	100	2000	3440	16.2	3.3	0.3	51.2	47.1	3.8	4.2	75.2
Ghana	270	2003	1148	25.4	14.2	2.1	29.4	21.8	23.2	5	28.2
Kenya	350	2003	1696	21.9	16.8	1.4	30.6	19.8	21.2	0.9	12.7
Malawi	160	2000	3136	18	2	2.9	49	25.4	13.4	2.7	27
Nigeria	300	2003	1535	36.1	29	2.2	38.5	28.7	21.8	3.7	41.6
Uganda	240	2000-2001	1796	20.8	6.2	1.4	38.6	22.5	15.1	3.8	21.9
Zambia	340	2001-2002	1897	25.6	15.6	1.6	46.8	28.2	11.2	3.7	12.1
The AMERICAS											
Bolivia	920	2003	3836	54	55.9	4.2	26.4	7.4	52.2	3.8	6.2
Colombia	1810	2000	1590	67.6	81.9	2.8	13.5	6.7	44.8	1.6	3.3
Guatemala	1750	1998-1999	1492	24.1	34.9	2.2	46.4	24.2	44.8	1.3	25.3
Haiti	440	2000	2044	26.8	11.6	0.9	21.9	16.8	24.4	1.5	28.9
Nicaragua	730	2001	2277	43.2	41.2	5.6	20.1	9.7	49.1	1.5	14.4
Peru	2020	2000??	4773	46.8	54.7	5.6	25.4	7.1	48.5	4.4	5.1

a. Based on number of children included in the samples used to build our model: country survey samples were much larger.
b. Based on total country sample (estimates provided by country at www..measuredhs.com).
c. Based on the limited samples used to build our model.
d. Based on total country survey sample aged 0-60 months, using >2 SD. Source: www.meaduredhs.com.
e. Based on total country survey sample, aged 0-60 months, using <2 SD. Source: www.meaduredhs.com
f. Median duration, in months.

The Country Samples

The seventeen country samples were based on Demographic and Health Surveys (Measure DHS+) studies conducted between 1998 and 2004 in Africa and the Americas. We selected those country surveys where anthropometric (height and weight) data was available for children aged 36-60 months and for their mothers. This provided a sample of 41,343 children. (We also originally planned to include a Bangladesh dataset as a South Asian case, which qualified based on these criteria, but the rate of child overweight was so very low that there was an insufficient basis to model risk. So we removed it from the study.) We then removed any cases where data were not available for all the variables of interest, and where the mother was more than three months pregnant (because it would affect the accuracy of their body weight measures), resulting in a final sample for analysis of 37,714. The DHS+ survey procedures are designed to be identical across countries: the standardized data collection procedures, including anthropometric measurement, interview schedules, and sampling frames can be found at the Marco International website (www.measuredhs.com). The basic sampling strategy is to produce a nationally-representative sample of households through a multiple-stage cluster design that begins with a selection of primary sampling units then randomly selects household within each (ORC Macro 1996). The primary participants in the surveys are all women aged 15-49 in the selected households, with all their children under five years then also targeted for anthropometric data collection. The sample size and collection dates for the included country surveys, along with some basic descriptive statistics, are presented in Table One.

The Variables and Model

A logistic regression model was employed to determine which of the predictors were significantly associated with the overweight status of the child. Because the relationships proved to vary substantially by region, we modeled each separately. In the final models for each region, all predictors and two-way interactions were considered; standard model-building techniques were employed to construct the "best" model, i.e., all predictors that remained significant ($p<0.01$) in the presence of other predictors, and there was no lack-of-fit as determined by the Hosmer-Lemeshow statistic. The dependent variable of interest was children's overweight status, which was derived from their weight-for-height measures, following World Health Organization recommendations (World Health Organization 1995). This was dichotomously coded based on whether weight-for-height was (1) or was not (0) more than two standard deviations (>2 SD) above the CDC growth reference curves derived from the NCHS/FELS/CDC reference populations. The predictor (independent) variables were:

A. Absolute household assets/wealth The DHS surveys collect categorical information on selected household amenities, such as water source, toilet type, housing materials, and ownership of durable goods. In other studies considering health-socioeconomic relationships using DHS data it has been proposed that these variables can, via scoring of items followed by a factor analysis, be used to create a "wealth index" that is a reasonable and usable proxy for expressing summary household variation in income and expenditure (Rutstein and

Johnson 2004, Filmer and Prichett 2001). Factor analysis is generally applicable for variables with values on a continuous scale, creating a smaller set of variables that are also on a continuous scale. Using a continuous variable in logistic regression is often leads to lack of fit because a very specific functional form for the probability curve is assumed. Categorical predictors are much less subject to lack of fit. Thus, we created an absolute household wealth variable with three levels: (1) bare floors, no electricity, and no flush toilet available (66.75% of the households of sample children in sub-Saharan Africa, 36.19% in the Americas and 5.47% in Northern Africa), (3) homes with flush toilet, electricity, finished floors, and either a television and/or a car (2.73% in sub-Saharan Africa, 21.07% in the Americas, and 60.20% in North Africa), and (2) all other households (i.e., the middle wealth category). This seemed a reasonable categorization, given that over half of children's households had essentially no basic services (that is, were classifiable as level 1). It is thus important to also note that being classified in the more affluent category in this sample in fact simply indicates a relative absence of poverty. The inclusion of a household social status variable (below) was intended to capture some different aspects of the variation in income beyond this.

B. Household social status. This variable was based on parental occupation and education. For the same reasons we outlined in considering household wealth, we found that creation of a standardized index based on factor analysis was not justifiable. Instead we identified six levels of the variable, which is feasible given the very large size of the sample. Paternal education and occupation tends to display far more heterogeneity than that of women (Fotso and Kuate-Defo 2005), so we built categories of social status with this in mind. The categorizations also recognize that many households, especially in the Americas, are female-headed, so that social status was assigned on the basis of either husband or wife's education and/or occupation (husband or dual-headed households) or wife's status alone (female headed-households). The six levels were:

[1] Professionals: Woman or her partner works in professional, technical, or managerial position or either has some tertiary education.
[2] Educated agricultural self-employed: Woman or her partner is self-employed in the agricultural sector, and either completed primary school.
[3] Educated service employed: Woman or her partner works in sales, clerical, skilled manual, or services industry, and at least one attended primary school.
[4] Educated unskilled laborer: Either woman or husband completed primary school and is employed as domestic, unskilled, or agricultural laborer.
[5] Unschooled agricultural self-employed: Woman or partner is agricultural self-employed and neither completed primary school.
[6] All others: unskilled labor or agricultural employment where either woman or partner did not attend school, or both the woman and partner are unemployed or the women does not have a partner and is unemployed.

C. Region. Separate logistic regression models were developed for each region. We categorized countries into the following regions: Northern Africa (Morocco and Egypt), Sub-Saharan Africa (including Ethiopia), and the Americas (including Haiti). See Table 1.

D. Urban-rural residence. This was entered into the model as a two-level variable based on women's de facto residence at the time she was interviewed. A strict urban/rural distinction can hide much heterogeneity in living conditions and so in related risk factors;

particularly there can be important differences between living in an urban core versus the peri-urban surrounds of large cities versus a rural town and the countryside (McDade and Adair 2001). For most of the cases (N=37,071) we had data further distinguished into four categories: capital city or city over one million people, small cities (population over 50,000), town, or the countryside. When we ran the model based on these cases, we found the results were essentially the same (i.e., the differences were between small town/countryside and small/large city). We thus completed the model using the two level urban/rural categories.

E. Mother's weight status. This was classified based on mother's body mass index (BMI = kg/m^2), whereby overweight was BMI of greater than 25, underweight was BMI of less than 18.5, and normal weight was anything between these. Overweight entered into the model as 3, normal weight as 2, and underweight as 1.

F. Mother's (a) age and (b) total number of other children in the household. Mother's risk of overweight tends to increase both with her age and parity, so these were also entered into the model. The number of other young children in the household (highly associated with maternal parity) should also directly impact the availability of food to the target child, especially under extreme poverty conditions. Mother's age was entered into the model in the following three categories: 15-24 years, 25-34 years, and 35-49 years. The number of siblings under age five currently living in the household was entered into the model in following categories: no other children, 1-2 others, or more than 2 others.

G. Child's breastfeeding history. This was based on mother's reports of age in months at weaning and was categorized as follows: target child was breastfed for 0-11 months, 12-35 months, or 36 or more months.

H. Growth stunting (very low height-for-age) has been observed to place developing country children at increased risk of overweight, especially in middle income nations (e.g., Popkin et al., 1996, Sawaya et al., 1998), so was included in the model. This was entered as a dummy variable categorized on the basis of whether the child was below two standard deviations (<-2 SD) of height-for-age (1) or not (0), based on comparison with reference curves derived from the NCHS/FELS/CDC reference populations. Of the total sample, 37.6 percent of children were classified as stunted: the range was from 14.5 percent in Colombia to almost 60 percent in Guatemala. Stunting was more common than overweight in all the country samples, although some children in all regions were affected by both (see Table 2).

I. Other Variables: We intended to enter country wealth (per-capita gross national income [GNI]) into our model. The per-capita GNI of countries in our sample ranged from $100 to $2,020 (Table 1), which is classified by the World Bank as very low to lower-middle income (World Bank 2005). We found that because most of the countries we are sampling are very low income or low income, the range was fairly narrow and the levels in wealth were distinguished mostly by region; region and country wealth are thus strong proxies for each other in the model. Children's gender was not a significant predictor of overweight status either as a main effect in combination with other variables; it was removed from the model. We also had available birth weight data for a subset of the children (N=14,847). Based on analyses of these children, we found low birth weight (<2500 grams) did not predict overweight status, and thus could reasonably be removed.

Table 2: Percentage of Children Stunted and Overweight by Region

	Overweight	Stunted	Both Overweight and Stunted		
			Total	Urban only	Rural only
Sub-Saharan Africa	1.7%	47.3%	.9%	0.5%	1.0%
Northern Africa	7.7%	17.4%	2.2%	2.2%	2.2%
The Americas	4.2%	31.5%	.9%	0.7%	1.1%

Results

We find that that absolute household wealth, household social status, mother's overweight status, breastfeeding, and urbanism have different effects for early childhood overweight risk in different world regions. Particularly, the patterns in the Americas are generally distinct from those observed for Africa, so the associations between household wealth, social status, and urban status in predicting child overweight risk need to be considered particular to each. The results of the logistic regression are presented in Tables 3, 4, and 5, differentiated by region. The first three figures illustrate the differences in risk of overweight for children in each region, and the plots show different patterns of overweight by some of the major predictors. North Africa has the highest prevalence of overweight children, and sub-Saharan Africa the lowest. Note that in reporting of the regional patterns, we do not discuss variables that proved to have no interactional or independent effect on children's overweight risk once other variables were taken into account.

Table 3: Results for the Model for Sub-Saharan Africa. All comparisons for household social status were made against the final category (6), which represented those households with the under or unemployed and least educated parents.

Predictor	Odds Ratio	95% Confidence Interval	p-value
Overweight mother	2.08	(1.47,2.96)	<0.0001
Household social status (category 1 versus 6)	2.46	(1.32,4.58)	0.0003
(2 versus 6)	2.64	(1.55,4.48)	
(3 versus 6)	3.09	(1.91,4.97)	
(4 versus 6)	1.81	(0.86,3.08)	
(5 versus 6)	2.13	(1.16,3.91)	
Stunted	1.46	(1.12,1.91)	0.0025

Table 4: Results for the Model for the Americas. All comparisons for household social status were made against the final category (6), which represented those households with the under or unemployed and least educated parents. Urban residence results are compared to rural with the same household wealth categories.

Predictor	Odds Ratio	95% Confidence Interval	p-value
Overweight mother	2.05	(1.72,2.44)	<0.0001
Small family	1.42	(1.18,1.70)	0.0001
Household Social Status (category 1 versus 6)	1.48	(1.10,2.00)	0.0003
(2 versus 6)	1.86	(1.35,2.55)	
(3 versus 6)	1.78	(1.26,2.52)	
(4 versus 6)	1.11	(0.80,1.54)	
(5 versus 6)	1.21	(0.89,1.64)	
Urban and lowest household wealth	0.30	(0.11,0.80)	<0.0001
Urban and middle household wealth	1.49	(1.17,1.90)	
Urban and highest household wealth	1.96	(1.52,2.53)	
Breastfeeding 12 -35 months	0.76	(0.64,0.91)	0.0025

Table 5. Results of the Model for Northern Africa. All noted categories report significantly different odds than their comparison categories.

Predictor	Odds Ratio	95% Confidence Interval	p-value
Overweight mother	2.44	(1.91,3.12)	<0.0001
Highest household wealth category	1.53	(1.20,1.95)	0.0007
Urban residence and breastfeed 12 -35 months	0.59	(0.45,0.76)	<0.0001
Stunted, rural residence	1.79	(1.34,2.38)	<0.0001
Stunted, urban residence	3.74	(2.63,5.33)	<0.0001

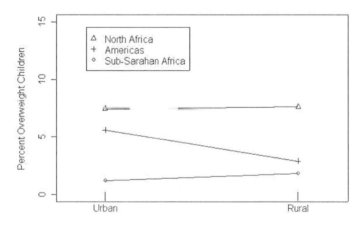

Figure 1. Percentage of Overweight Children by Urban/Rural Residence by Developing Country Region. The lines on all the figures are for visual purposes only.

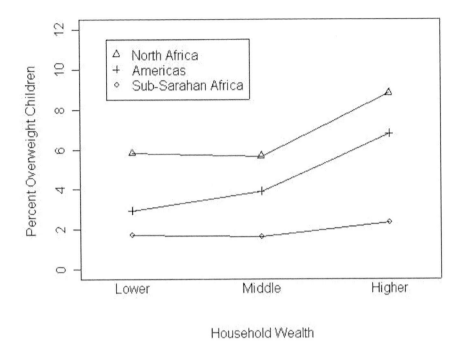

Figure 2. Percentage of Overweight Children by Level of Household Wealth by Developing Country Region.

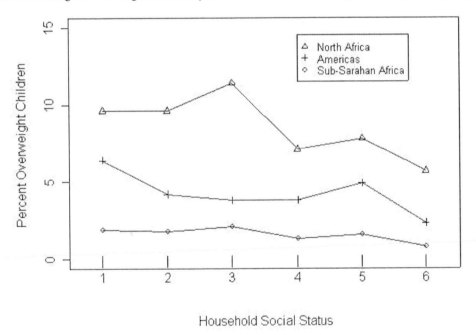

Figure 3. Percentage of Overweight Children by Household Status (based on parental education and occupation) by Developing Country Region. Note: the categories are not necessarily ordered, although group 6 is definitely the lowest status (mostly unemployed or unskilled with no education) and group 1 the highest (at least one parent is professionally employed and better educated).

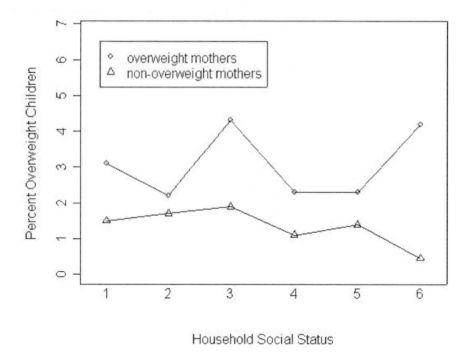

Figure 4. Percentage of Overweight Children by Household Social Status and Overweight Status of Mothers in Sub-Saharan Africa. See the Figure 3 legend regarding the social status categories.

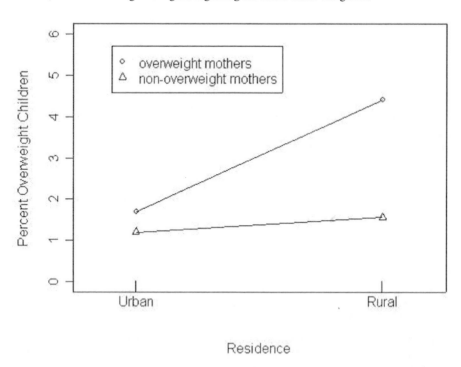

Figure 5. Percentage of Overweight Children by Household Wealth Level and Overweight Status of Mothers in Sub-Saharan Africa.

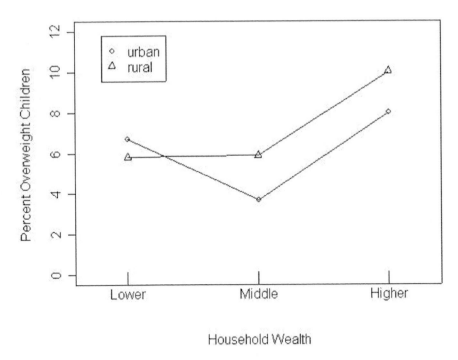

Figure 6. Percentage of Overweight Children by Urban/Rural Residence and Overweight Status of Mothers in Northern Africa.

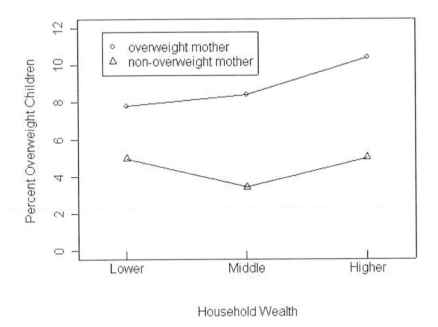

Figure 7. Percentage of Overweight Children by Household Wealth and Overweight Status of Mothers in Northern Africa.

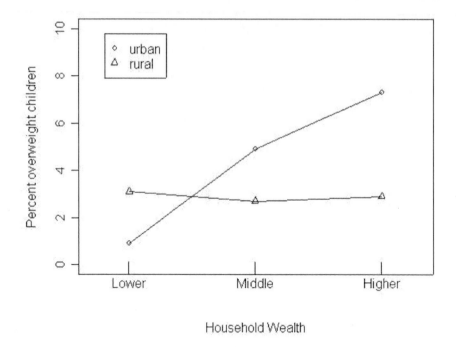

Figure 8. Percentage of Overweight Children by Rural/Urban Residence and Household Wealth in the Americas.

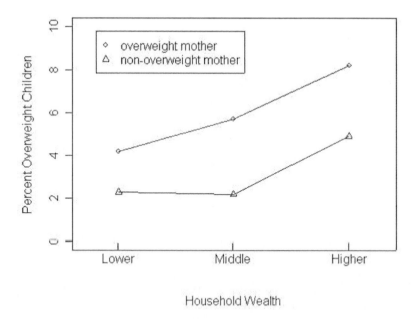

Figure 9. Percentage of Overweight Children by Household Wealth and Overweight Status of Mothers in the Americas.

The Pattern in Sub-Saharan Africa. Rates of overweight in early childhood are very low in all the sub-Saharan African countries. The strongest predictor of child overweight is the overweight status of the mother. The logistic regression model results (see Table 3) indicate that the odds of child being overweight is slightly more than two times higher for children of overweight mothers compared to children of mothers who are not overweight. Of course, this may be a surrogate for household social and economic conditions, as presumably these are predictors of mother's overweight, but the number of overweight children in the sample is too small to be able to detect any more precision if this is the case. Stunting (representing prior chronic under-nutrition) also places Sub-Saharan children at increased risk: stunted children have odds of overweight about 1.5 times higher than those who are not stunted. The household social status is a significant predictor of child overweight, with the first five categories having higher odds of overweight than the sixth. In particular, categories 1-3 have odds of overweight at least 2.5 times higher than category 6, and these differences are highly statistically significant.

The Pattern in Northern Africa. The strongest predictors of child overweight in the two Northern African countries are the mother's overweight status, whether the child's home is urban or rural, and whether the child is stunted. We see a 144% increased odds of child overweight if the mother is overweight. The type of residence strongly interacts with stunting and with long breastfeeding. The effect of stunting on risk of overweight in North Africa is different for urban and rural children. Urban children's odds of overweight are 3.7 times higher if the child is stunted, and rural children's odds of overweight are 1.8 times higher if the child is stunted. If the child is not stunted, the risk of overweight is not significantly different by type of residence (Figure 6). If the child's household wealth is of the highest level, the child has 53% larger odds of overweight compared with children in the less affluent categories (Figure 7). Finally, if an urban child is breastfed for at least 36 months, the odds of overweight are 41% lower than otherwise, but there is no breastfeeding effect for rural children.

The Pattern in the Americas. There are many highly significant predictors of child overweight in the developing countries of the Americas. Social and wealth status of the child's household are strong predictors, with wealth interacting strongly with the urban/rural location of the residence. The interaction effect is illustrated in Figure 8: we see that for urban children there is a strong relationship between family wealth and risk of overweight, with children from more wealthy families showing higher probability of overweight. For rural children, the risk of overweight does not vary significantly with family wealth. The logistic regression output (Table 5) allows us to quantify the risk of overweight by family affluence and on the urban/rural residence indicator. Because risk of overweight for rural children does not vary significantly by family affluence, we compare urban affluence groups to rural. Children in the urban, poor group have 70% lower odds of being overweight compared with rural children, while children in the urban, middle wealth group have 49% higher odds, and children in the urban, affluent group have 96% higher odds, compared with rural children. Family size is a predictor of overweight, with children from smaller families having larger risk of overweight. Specifically, the odds of overweight for children with zero or one sibling is 42% higher than the odds for children with 2 or more siblings. Children with parents with the lowest education and occupational status (category 6) have the lowest risk of child overweight. Compared to this group, children in groups 1, 2 and 3 (children of professionals, and educated skilled workers and self-employed agriculturalists) have significantly higher risk: odds are 48%, 86%, and 78% higher, respectively. The risk of overweight for children in groups 4 and 5 (children with less skilled and less educated parents) do not vary significantly from those in group 6 (the least skilled and educated). Breastfeeding duration is found to be associated with risk of overweight, with children that were breastfed for a year or more having 24% smaller odds of overweight, compared with children that were breastfed less than a year. Notably, stunting did not predict any increased risk of a child also being overweight.

The "Global" Pattern. Placing the regional patterns in the context of each other we find that controlling for household social and wealth categories, it is rural children in the Africas who are more at risk of overweight (although the effect is not a particularly strong one) and it is urban children in the Americas. In the Americas, rural children show no difference in overweight risk based on household wealth, but in urban contexts this has a substantial effect – whereby, the more affluent the household, the higher the risk of being overweight. In all regions, children are more likely to be overweight if their mothers are overweight: in sum, it is the strongest predictor of all. In sub-Saharan Africa, where almost all the households are in the very poor category, mother's overweight status is the only very strong significant predictor of child overweight risk. In this situation, it is likely that mother's nutritional status is a proxy for some aspects of household wealth and/or social status factors we have failed to capture, but we cannot observe this variation because the number of overweight children is so few (even with this very large sample size). For Northern Africa, the predictors are much the same, but there is a much higher prevalence rate of child overweight overall, and children from most affluent households are more likely to be overweight, controlling for differences in household social status. In the Americas, we observe that household social status exerts an influence on child overweight risk that was independent of household wealth (and vice versa). That is, in rural households risk of being overweight varies by social status but not with wealth, but in urban households it varies by wealth even once social status is controlled for. Breastfeeding only predicts overweight status in the Americas, in which case it is protective: where children are breastfed for at least one year they are more likely to be overweight once household location, wealth, and social status are taken into account. Overweight risk was also higher where there are not other young siblings in the home. The finding that child stunting increases the risk of child overweight, at least in Africa, and that this effect is strongest in urban settings (at least in Northern Africa, where child obesity was much more common), is consistent with what has been observed in the few studies that have been done in middle income developing countries (e.g., Sawaya 1998). It is particularly important to note that the risk of child stunting in this sample, if we run it as an outcome variable, is very strongly predicted by household wealth: meaning that it is in the poorer households that we can observe stunting emerging as a critical predictor of childhood overweight risk, even at these very young ages.

Conclusion

Once we control for social status, household wealth, and such factors as children's prior under-nutrition (stunting) and maternal weight status, urbanism plays a limited role in constructing children's risk of being overweight. Household economics do have a significant role: the wealthiest families in Africa, and the wealthiest urban families in the Americas have the greatest risk of early childhood overweight. As low income countries continue a trajectory of economic growth, however, rising rates of child overweight are predicted. Stunting, which is much more common in this sample of developing country children than overweight, does place children in both sub-Saharan and North Africa (but not the Americas) at an elevated risk of potentially unhealthy levels of weight gain, even when you control for such other factors as income and even mother's weight status. However, we also detected an interaction between urban residence and stunting, in that children in urban settings have considerably greater risk of being overweight if they are stunted, compared to their stunted rural counterparts. Overall, however, it is the weight status of mothers independent of all these other factors that appears to best predict young children's risk of being overweight.

What implications does this have as we consider emerging risks for child obesity in the poorer developing countries? First, the model suggests that women's obesity rates are the harbingers of an emerging risk of early childhood obesity, indicating both the possible scale and where it is most likely to occur. It also suggests that increases the absolute wealth of households among the very poor, tied to the economic development of countries, will increase children's risk of over-nutrition, possibly very rapidly. Given the prevalence of stunting in many of these populations and ongoing urbanization, it means that childhood overweight and stunting will likely increasingly course together to amplify risk – perhaps most especially in Africa. Thus, child obesity may well emerge as an embodiment of relative poverty, just as stunting reflects the manifestation of absolute poverty. Moreover, the goal of a healthy 'normal' weight could prove as elusive and transitory for many of the world's least affluent, children, just as it now increasingly appears to be for their mothers.

References

Ball, K. & Crawford, D., (2005). Socioeconomic status and weight change in adults: A review. *Social Science and Medicine* 60, 1817-2010.

Brewis, A. (2003). Biocultural aspects of obesity in young Mexican schoolchildren, *American Journal of Human Biology,* 15, 446-60.

Brody, J. (2002). The global epidemic of childhood obesity: Poverty, urbanization, and the nutrition transition. *Nutrition Bytes,* 8, Article 1. http://repositories.cdlib.org/uclabiolchem/nutritionbytes/vol8/iss2/art1

Caballero, B. (2005). A nutrition paradox – underweight and obesity in developing countries. *New England Journal of Medicine* 352:1514-1516.

Crooks, D. L. (1995). American children at risk: Poverty and its consequences for growth, health and school achievement. *Yearbook of Physical Anthropology,* 38,57-86.

Darnton-Hill, I. & Coyne, E. (1998). Feast and famine: Socioeconomic disparities in global nutrition and health. *Public Health Nutrition* 1, 23-31.

Filmer, D. & Pritchett, L. (2001). Estimating wealth effects without expenditure data—or tears: An application to educational enrollments in states in India. *Demography,* 38, 115-132.

Fotso, J-C., & Kuate-Defo, B., (2005). Household and community socioeconomic influences on early childhood malnutrition in Africa. *Journal of Biosocial Science,* 38, 289-313.

Hedley, A., Ogden, C., Johnson, C., Carroll, M., Curtin, M., & Flegal, K. (2004). Prevlance of overweight and obesity among US children, adolescents, and adults, 1999-2002. *Journal of the American Medical Association*, 291, 2847-2850.

Jebb, S., Rennie, K., & Cole, T. (2003). Prevlance of overweight and obesity among young people in Great Britain. *Public Health Nutrition*, 7, 461-465.

Livingstone, M. (2000). Epidemiology of childhood obesity in Europe. *European Journal of Pediatrics*, 195, S14-S34.

McDade, T. & Adair, L. (2001). Defining the "urban" in urbanization and health: A factor analysis approach. *Social Science & Medicine*, 53, 55-70.

ORC Macro (1996) Sampling Manual. *DHS-III Basic Documentation*, No. 6. Calverton, Maryland.

Magarey A.M., Daniels L.A., & Boulton T.J.(2001) Prevalence of overweight and obesity in Australian children and adolescents: reassessment of 1985 and 1995 data against new standard international definitions. *Medical Journal of Australia,* 174, 561-565.

Martorell, R., Kettel Khan, L., Hughes, M.L., & Grummer-Strawn, L.M. (2000). Overweight and obesity in preschool children from developing countries. *International Journal of Obesity*, 24, 959-967.

Mendez, M.A. & Popkin, B. (2004). Globalization, urbanization, and nutritional change in the developing world. *Journal of Agricultural and Development Economics*. 1, 220-241.

Monteiro, C., Conde, W.L., Lu, B., & Popkin, B. (2004a). Obesity and inequities in health in the developing world. *International Journal of Obesity*, 28, 1181-1186.

Monteiro, C., Moura, E., Conde, W., & Popkin, B. (2004b). Socioeconomic status and obesity in adult populations of developing countries: a review. *Bulletin of the World Health Organization*, 82, 940-946.

Ogden, C., Troiano, R., Briefel, R., Kuczmarski, R., Flegal, K., & Johnson, C. (1997). Prevalence of overweight among preschool children in the United States, 1971 through 1994. *Pediatrics* 99, e1.

Popkin, B. (2004). The nutrition transition: Patterns of Change. *Nutrition Reviews*, 62, S140-143.

Popkin, B., Richards, M., & Monteiro, C. (1996). Stunting is associated with overweight in children of four nations that are undergoing the nutrition transition. *Journal of Nutrition*, 126, 3009-3016.

Rutstein, S., & Johnson, K. (2004) The DHS Wealth Index. *DHS Comparative Reports* No. 6. ORC Macro, Calverton, Maryland.

Sawaya, A.L., Grillo, L., Verreschi, I., da Silva, A.C., & Roberts, S. (1998). Mild stunting is associated with higher susceptibility to the effects of high fat diets: Studies in a shantytown population in São Paulo, Brazil. *Journal of Nutrition* 128, 415S-420S.

Shetty P. S. (1999). Obesity in children in developing societies: indicator of economic progress or a prelude to a health disaster [editorial]. *Indian Pediatrics*, 36, 11-15.

Sobal, J. & Stunkard, A.J. (1989). Socioeconomic status and obesity: A review of the literature. *Psychological Bulletin*, 105, 260-275.

Wang, Y. (2001), Cross-national comparison of childhood obesity: The epidemic and the relationship between obesity and socioeconomic status. *International Journal of epidemiology*, 30, 1129-1136.

World Bank, (2005). World Development Indicators 2005. World Bank, Washington, DC.

World Health Organization [WHO], (1998). Obesity: preventing and managing the global epidemic. *Report of a WHO consultation on obesity*. WHO, Geneva.

World Health Organization [WHO], (1995). Physical Status: The Use and Interpretation of Anthropometry. *WHO Technical Report* 1995:854.

Chapter Sources

The following chapters have been previously published:

Chapter 11 – A version of these chapters was also published in *Nutritional Education*, edited by Ida R. Laidyth, published by Nova Science Publishers, Inc. It was submitted for appropriate modifications in an effort to encourage wider dissemination of research.

Chapter 12 – A version of these chapters was also published in *Global Dimensions of Childhood Obesity*, edited by Richard K. Flamenbaum, published by Nova Science Publishers, Inc. It was submitted for appropriate modifications in an effort to encourage wider dissemination of research.

Index

C

D

E

F

G